Applications of NMR Spectroscopy
(Volume 7)

Edited by

Atta-ur-Rahman, *FRS*

Honorary Life Fellow,
Kings College, University of Cambridge, Cambridge, UK

&

M. Iqbal Choudhary

H.E.J. Research Institute of Chemistry, International Center for Chemical
and Biological Sciences, University of Karachi, Karachi, Pakistan

Applications of NMR Spectroscopy

Volume # 7

Editors: Atta-ur-Rahman and M. Iqbal Choudhary

ISSN (Online): 2405-4682

ISSN (Print): 2405-4674

ISBN (Online): 978-1-68108-641-5

ISBN (Print): 978-1-68108-642-2

©2019, Bentham eBooks imprint.

Published by Bentham Science Publishers – Sharjah, UAE. All Rights Reserved.

need for a court order if at any point you breach any terms of this License Agreement. In no event will any delay or failure by Bentham Science Publishers in enforcing your compliance with this License Agreement constitute a waiver of any of its rights.

3. You acknowledge that you have read this License Agreement, and agree to be bound by its terms and conditions. To the extent that any other terms and conditions presented on any website of Bentham Science Publishers conflict with, or are inconsistent with, the terms and conditions set out in this License Agreement, you acknowledge that the terms and conditions set out in this License Agreement shall prevail.

Bentham Science Publishers Ltd.
Executive Suite Y - 2
PO Box 7917, Saif Zone
Sharjah, U.A.E.
Email: subscriptions@benthamscience.net

BENTHAM SCIENCE

CONTENTS

PREFACE

The applications of NMR spectroscopy are now expanding to new and diverse disciplines. Volume 7 of the book series, *"Applications of NMR Spectroscopy"*, is an excellent compilation of five scholarly written articles, each focusing on some of these diverse applications of NMR with a futuristic perspective.

Habenstein *et al.* have contributed a review on the use of solid-state NMR spectroscopy for structural studies of self-assembled macromolecular complexes. They have demonstrated the potential of HR-MAS NMR spectroscopy in obtaining the atomic level resolution of such large protein complexes. Cox *et al.* have reviewed the recent literature on the applications of ^{1}H-NMR-based metabolomics in biofluids for early diagnosis of liver diseases, and to identify microbial co-metabolites. Guillen *et al.* present an article on NMR-based conditions of the digestive changes in various lipids and fats, demonstrating interesting biochemical changes taking place in these essential nutrients.

On the same lines, Iadarola *et al.* have used NMR for the identification of chemical constituents in acute and chronic lung disorders, that can serve as biomarkers of the disease. Kaushik *et al.* have contributed a comprehensive review on ^{1}H-,^{13}C- and ^{31}P-NMR-based structure determination of DNA topologies, and dynamic changes in their structures as a result of interactions with various ligands. The review contributed by Kumar *et al.* focusses on the possible applications of NMR in disease metabolomics, particularly in the early diagnosis of cancer therapy through sensitive, reproducible, and non-destructive analysis of diverse metabolites.

We are grateful to Ms. Fariya Zulfiqar (Manager Publications), Mr. Shehzad Naqvi (Editorial Manager Publications), Mr. Mahmood Alam (Director Publications) and other members of Bentham Science Publishers for the excellent coordination with the authors, and timely compilation of the volume for the readers.

Prof. Dr. Atta-ur-Rahman, FRS
Honorary Life Fellow,
Kings College,
University of Cambridge,
Cambridge,
UK

&

Prof. Dr. M. Iqbal Choudhary
H.E.J. Research Institute of Chemistry,
International Center for Chemical and Biological Sciences,
University of Karachi,
Pakistan

List of Contributors

Antoine Loquet
Institute of Chemistry & Biology of Membranes & Nanoobjects (CBMN UMR5248), CNRS, University of Bordeaux, Institut Européen de Chimie et Biologie, All. Geoffroy Saint-Hilaire, Pessac, France

Birgit Habenstein
Institute of Chemistry & Biology of Membranes & Nanoobjects (CBMN UMR5248), CNRS, University of Bordeaux, Institut Européen de Chimie et Biologie, All. Geoffroy Saint-Hilaire, Pessac, France

Bárbara Nieva-Echevarría
Food Technology, Faculty of Pharmacy, Lascaray Research Center, University of the Basque Country (UPV/EHU), Paseo de la Universidad n° 7, 01006 Vitoria-Gasteiz, Spain

Cristina Airoldi
Department of Biotechnologies and Biosciences, University of Milano-Bicocca, Piazza della Scienza 2, Milano, Italy

Carlotta Ciaramelli
Department of Biotechnologies and Biosciences, University of Milano-Bicocca, Piazza della Scienza 2, Milano, Italy

Dinesh Kumar
School of Chemical Sciences, Central University of Gujrat, Gandhinagar 382030, India

Encarnación Goicoechea
Food Technology, Faculty of Pharmacy, Lascaray Research Center, University of the Basque Country (UPV/EHU), Paseo de la Universidad n° 7, 01006 Vitoria-Gasteiz, Spain

I. Jane Cox
Institute of Hepatology, London, Foundation for Liver Research, 111 Coldharbour Lane, London SE5 9NT, United Kingdom
Faculty of Life Sciences & Medicine, King's College London, London SE5 9RS, United Kingdom

James Tolchard
Institute of Chemistry & Biology of Membranes & Nanoobjects (CBMN UMR5248), CNRS, University of Bordeaux, Institut Européen de Chimie et Biologie, All. Geoffroy Saint-Hilaire, Pessac, France

Komal Mehra
Nucleic Acids Research Laboratory, Department of Chemistry, University of Delhi, Delhi, India

Mark J.W. McPhail
Department of Inflammation Biology, School of Immunology and Microbial Sciences, Faculty of Life Sciences and Medicine, Institute of Liver Sciences, King's College London, London SE5 9RS, United Kingdom

María D. Guillén
Food Technology, Faculty of Pharmacy, Lascaray Research Center, University of the Basque Country (UPV/EHU), Paseo de la Universidad n° 7, 01006 Vitoria-Gasteiz, Spain

Mahima Kaushik
Cluster Innovation Centre, University of Delhi, Delhi, India
Nucleic Acids Research Laboratory, Department of Chemistry, University of Delhi, Delhi, India

Manjula Nair
HBMSU, Academic City, Dubai, UAE

Neetu Talreja
Department of Bio-nanotechnology, Gachon University, Gyeonggi-do, South Korea

Paolo Iadarola
Department of Biology and Biotechnologies "L.Spallanzani", University of Pavia, *Via* A.Ferrata 9, Pavia, Italy

Roger Williams	Institute of Hepatology, London, Foundation for Liver Research, 111 Coldharbour Lane, London SE5 9NT, United Kingdom Faculty of Life Sciences & Medicine, King's College London, London SE5 9RS, United Kingdom
Simona Viglio	Department of Molecular Medicine, University of Pavia, *Via* Taramelli 3, Pavia, Italy
Shrikant Kukreti	Nucleic Acids Research Laboratory, Department of Chemistry, University of Delhi, Delhi, India
Sonia Khurana	Nucleic Acids Research Laboratory, Department of Chemistry, University of Delhi, Delhi, India
Swati Chaudhary	Nucleic Acids Research Laboratory, Department of Chemistry, University of Delhi, Delhi, India

CHAPTER 1

Atomic Structural Investigations of Self-Assembled Protein Complexes by Solid-State NMR

Antoine Loquet*, James Tolchard and Birgit Habenstein*

Institute of Chemistry & Biology of Membranes & Nanoobjects (CBMN UMR5248), CNRS, University of Bordeaux, Institut Européen de Chimie et Biologie, All. Geoffroy Saint-Hilaire, 33600 Pessac, France

Abstract: Protein self-assemblies play essential roles in many biological processes ranging from bacterial and viral infections to basic cellular functions. They can be found in a wide range of supramolecular architectures, in homomeric and heteromeric forms, and often in symmetric arrangements. The biological function will then be dictated by the structure of the assembled object rather than by the subunits. Atomic-resolution structural investigations of protein assemblies can be tedious because of their size, their insolubility, and often their non-crystallinity. Solid-state NMR (ssNMR) is a powerful technique used to obtain high-resolution structural models of these complex assemblies and to study their assembly processes and interactions at the atomic level. Unrestricted by object size or solubility, ssNMR can be applied to study the structures and interactions of macromolecular assemblies such as proteins in a membrane environment, protein filaments, pores, fibrils or oligomeric species. This chapter focusses on the established methods and recent advances in magic angle spinning (MAS) ssNMR for the detection of structural restraints in macromolecular protein assemblies and the determination of their atomic-resolution models. We will review different ^{13}C and ^{15}N isotope labelling approaches necessary to detect and differentiate intra- and intermolecular distance restraints that define the protein subunit structure and their relevance in the context of symmetric assemblies. The collection, and interpretation of, structural restraints in protein assemblies by ssNMR will be discussed. We will also introduce the recent developments in ultra-fast MAS ssNMR to study and determine atomic structures of sub-milligram quantities of molecular assemblies using proton detection. Finally, our aim is to also illustrate the complementarity of ssNMR to other techniques in structural biology such as solution-state NMR, mass-per-length scanning transmission electron microscopy (STEM) measurements and cryo-electron microscopy.

* **Corresponding authors:** Institute of Chemistry & Biology of Membranes & Nanoobjects (CBMN UMR5248), CNRS, University of Bordeaux, Institut Européen de Chimie et Biologie, All. Geoffroy Saint-Hilaire, 33600 Pessac, France; Tel: +33 (0)5 40 00 30 29; E-mails: b.habenstein@iecb.u-bordeaux.fr; a.loquet@iecb.u-bordeaux.fr

Atta-ur-Rahman & M. Iqbal Choudhary (Eds.)

Keywords: Amyloid fibrils, Bacterial filaments, Macromolecular assemblies, Magic angle spinning, Nuclear magnetic resonance, Protein assembly, Protein complexes, Self-assemblies, Solid-state NMR, Structural biology, Structure determination.

INTRODUCTION

Self-assembled protein complexes occur in a variety of biological processes ranging from host-pathogen interactions to the pathogenesis of neurodegenerative disorders. Bacterial flagella, bacterial secretion systems, viral capsids, microtubules and ATP synthase are well-known examples of biomolecular assemblies. Nevertheless, three-dimensional atomic structures of these assemblies are underrepresented in databanks, since their supramolecular and non-crystalline state engenders an important challenge for structure determination techniques. Over the last decade solid-state NMR (ssNMR) has evolved into a powerful technique for the atomic-level structural assessment of fibrils and aggregates [1 - 3], bacterial appendages [4], complex protein assemblies [5 - 9], proteins in membrane environments [10 - 12], and dynamics [13 - 16] and interactions in assembled protein complexes. ssNMR has the advantage of being a non-destructive technique that can be applied to the relevant, assembled state of proteins without the need for subunit truncation or mutation to achieve solubility or a crystalline matrix. Notably, methodological and technological advances have propelled ssNMR to the forefront in the atomic studies of biomolecular assemblies. The implementation of selective labelling strategies and the growing diversity of available ssNMR experiments extend its application to an increasing number of problems. Recent pulse schemes allow for more efficient polarization transfer and specific transfers between selected atom types. Additionally, recent commercial proposals for higher magnetic field strengths (> 1 GHz) and faster magic angle spinning (> 100 kHz) (respectively reviewed here [17, 18]) will serve to increase the technological potential further still towards more sophisticated applications. In this chapter we will focus on state-of-the-art MAS ssNMR methods and applications to study structures of biomolecular self-assemblies at the level of atomic resolution.

Structure determination of supramolecular assemblies by ssNMR has many similarities to solution NMR structural studies, such as the requirement for isotope-labelled protein expression and purification, NMR experimentation, NMR resonance assignment, the collection of long-range restraints, and subsequent structural modelling as demonstrated by the workflow in Fig. (**1**). Nonetheless, the ssNMR structure assessment process does require some additional and substantially modified tasks. The protein subunits are expressed in ^{13}C, ^{15}N-labeled minimal medium, where the sole carbon and nitrogen sources are replaced with their ^{13}C- and ^{15}N-labeled analogues. Selective ^{13}C-labeling of the precursors such as [1,3-^{13}C]-Glycerol (Gly), [2-^{13}C]-Gly, [1-^{13}C]-Glucose (Glc) or [2-^{13}C]-Glc results in the production of selectively labelled protein subunits with beneficial spectroscopic properties. The application of strategic isotope labelling schemes then facilitates resonance assignment and the detection and identification of long-range contacts as a result of the reduced spectral crowding [19] and enhanced polarization transfer due to reduced dipolar truncation. In contrast to solution NMR, after purification, the protein subunits self-assemble into a supramolecular architecture and are then recovered by centrifugation. A typical protocol of ssNMR sample preparation is described in [20].

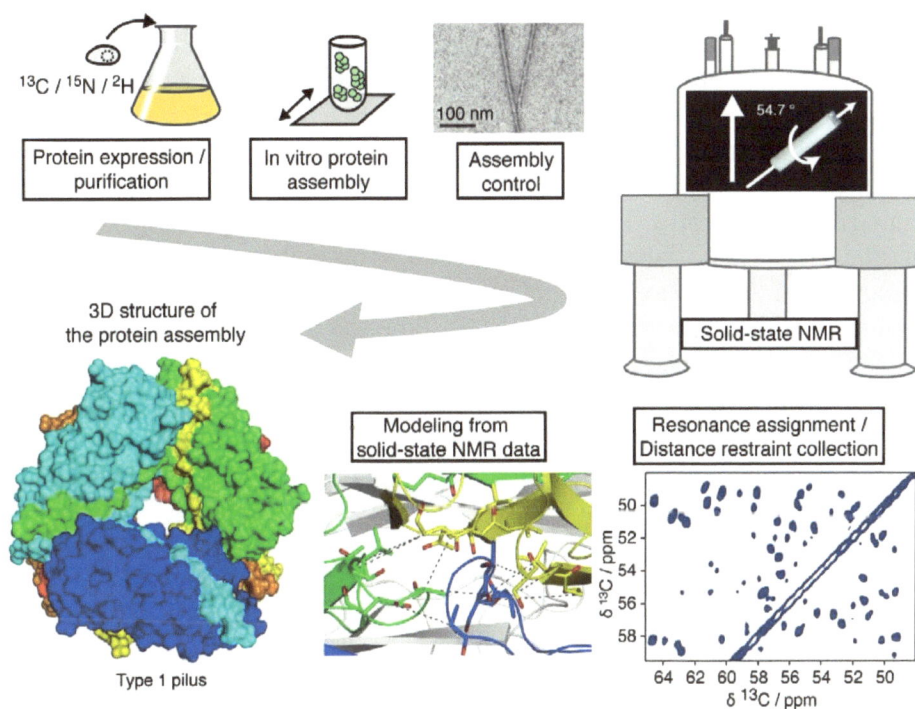

Fig. (1). The typical MAS ssNMR workflow for three-dimensional structure determination of protein assemblies.

Despite the rapid progress in ssNMR methodology, typical structural studies of self-assembled protein complexes still require more time for NMR experiment acquisition and spectral analysis than in solution NMR. Most current approaches rely on ^{13}C- and ^{15}N- detection because the detection of ^{1}H nuclei is only just becoming accessible, mainly with the advent of fast MAS ($\geq$ 60 kHz) ssNMR technology in combination with complete [21 - 23] or partial [24] deuteration [25], or ultra-fast MAS (usually >100kHz) with fully protonated protein systems [26 - 31], as briefly discussed at the end of the section. A detailed review from Pintacuda and co-workers nicely visualise the benefits in terms of magnetic field strength, increasing MAS speeds and partial deuteration [32]. The first step in the ssNMR analysis process is the establishment of ssNMR spectral fingerprints, followed by the NMR resonance assignment. This step consists of connecting each NMR signal to its respective nucleus. The assignment provides first insights into the assembled protein structure such as the secondary structure per residue, global structural order and structure rigidity. After ssNMR resonance assignment, the next step concerns the detection of close distance proximities (called long-range restraints, corresponding to a distance range of ~0.2-1 nm). In a supramolecular assembly this needs to be conceived in such a way so as to discriminate between intra- and intermolecular proximities, as illustrated in Fig. (**2**). In other words, distance restraints need to be collected to define (1) the 3D fold of the protein subunits and (2) their supramolecular arrangement in the assembled architecture. For this purpose, prior to the assembly process, equimolar mixtures of proteins with two different isotopic labelling schemes can be prepared, for example as illustrated in Fig. (**2B-C**) for (50/50) [^{15}N]/[^{13}C]-labelling [33] or (50/50) [1-^{13}C]-/[2-^{13}C]-Glc labelling [34], respectively. In a mixed 50/50 [^{15}N]/[^{13}C]-labelled supramolecular assembly, all signals visible in a two-dimensional ^{15}N-^{13}C experiment arise from intermolecular contacts (Fig. **2B**). Equally, in a (50/50) [1-^{13}C]-/[2-^{13}C]-Glc labelled sample, all ^{13}C-^{13}C contacts arising from two carbons in distinct positions, that are each only labelled in one of the ^{13}C-labelling schemes, are certainly intermolecular. Furthermore, a diluted sample (mixture of ^{13}C-labelled protein subunits in a unlabelled background) can be helpful in which intermolecular ^{13}C-^{13}C contacts disappear (or are strongly reduced in intensity) whereas intramolecular contacts should persist (Fig. **2D**).

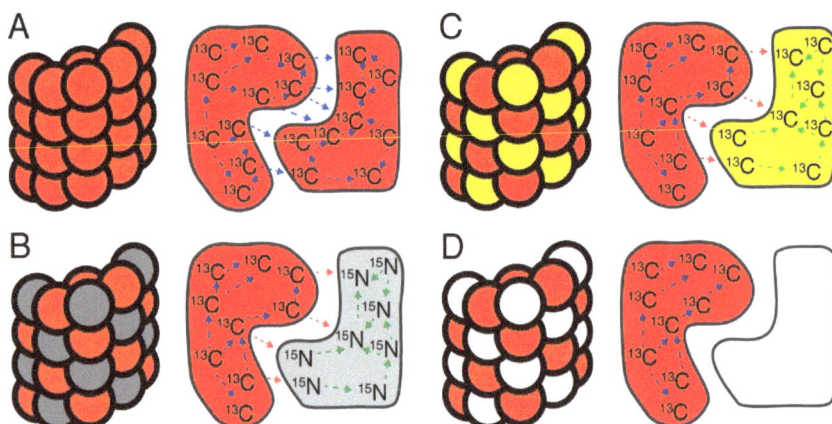

Fig. (2). Deriving intra- and intermolecular atomic distances from ssNMR with mixed isotopic labelling schemes. (**A**) Uniform ^{13}C labelling gives rise to NMR correlations for all proximal ^{13}C nuclei (intra- and intermolecular). Mixed (50/50) [^{15}N-/^{13}C]-labelled (**B**) and [1-^{13}C]-Glc/[2-^{13}C]-Glc-labelled (**C**) samples give rise to intermolecular contacts due to the obligatory polarization transfer steps. (**D**) Diluted samples (^{13}C-labelled subunits in an unlabelled background) promote the observation of intramolecular contacts.

From the ssNMR data, including intra- and intermolecular distance restraints, the molecular structure and its assembly architecture can be derived. Dihedral angles, as predicted from the NMR resonances (software TALOS+ [35] or PREDITOR [36]), identified long-range contacts, as well as additional structural data including mass-per-length measurements, electron microscopy density maps or mutagenesis data can be integrated as structural restraints into modelling software such as XPLOR-NIH [37], ARIA [38] or UNIO [39]. For a typical structure determination protocol of a supramolecular assembly please refer to [20].

An important aspect in the structure determination of complex assemblies concerns the symmetry of the supramolecular architecture. To illustrate the diversity of architecture symmetries, Fig. (**3**) shows the subunit arrangement of four structures that have been determined based on ssNMR. In bacterial appendages such as bacterial pili and secretion system needles, the subunits are commonly organized in a helical arrangement (Fig. **3A** and **B** [40, 41]) that, in some cases, allow for the accommodation of an interior lumen through which effector molecules might pass. In the particular case of helical assemblies, the identification of different supramolecular interfaces (*e.g.* between (i)-(i+1) and (i)-(i+2) in Fig. **3A**) between subunits is required. A helical subunit arrangement in the antiviral response protein MAVS [42] also serves to transduce the signal in mammalian innate immunity (Fig. **3C**). In amyloid fibrils, the supramolecular arrangement between subunits is often in a parallel or anti-parallel stacking (as illustrated in Fig. **3D**).

Inspired by solution NMR approaches, developments in ssNMR sample preparation and hardware design have started to allow for the direct detection of ^1H nuclei. The primary advantage of ^1H detection is that, with a fundamentally larger gyromagnetic ratio, ^1H nuclei are more sensitive to detection, which can act to facilitate the observation of dilute species, and/or serve to reduce experiment acquisition times. Moreover, the naturally abundant ^1H hydrogen isotope (99.98%) is readily NMR-active (nuclear spin ½) and constitutes a promising alternative to the expensive and complex sample preparation typically required for biological samples that are ^{13}C- and ^{15}N-labelled. In addition, the general ubiquity of ^1H makes it an extremely rich reporter for chemical environment throughout all biomolecules, and particularly so at inter- and intramolecular interfaces. However, the otherwise beneficially large gyromagnetic ratio of ^1H is also responsible for strong inherent homonuclear dipolar couplings, which low to moderate MAS speeds (<20 kHz) are unable to sufficiently diminish for the acquisition of well-resolved ^1H spectra. Nonetheless, to benefit from the aforementioned advantages direct ^1H-detection provides, two principal methodologies have so far been developed, with few exceptions such as dipolar filtering-based experiments [45]. The first method is based upon the dilution of protonated hydrogen sites to reduce the network of strong ^{1}H-^{1}H couplings. The dilution is performed by exchanging protons (^{1}H) for deuterons (^{2}H), through biochemical procedures such as back-exchange using a deuterated protein sample in protonated water. Whether the addition of ^{2}H groups is done specifically, randomly or completely (with some form of subsequent ^{1}H back-exchange) (see: isotopic labelling), the reduced ^{1}H concentration acts to reduce the dominance of ^{1}H-^{1}H dipolar couplings and can yield well-resolved ^{1}H-detected spectra [21 - 24]. The second approach relies on recent ssNMR probes that can attain MAS frequencies >100 kHz [27, 28, 46, 47], which are becoming increasingly available. So-called 'ultra-fast' MAS ssNMR is a very attractive prospect, as deuteration is no longer a prerequisite and can allow for markedly improved ^{1}H spectral resolution, ^{1}H assignment and the detection of long-range restraints.

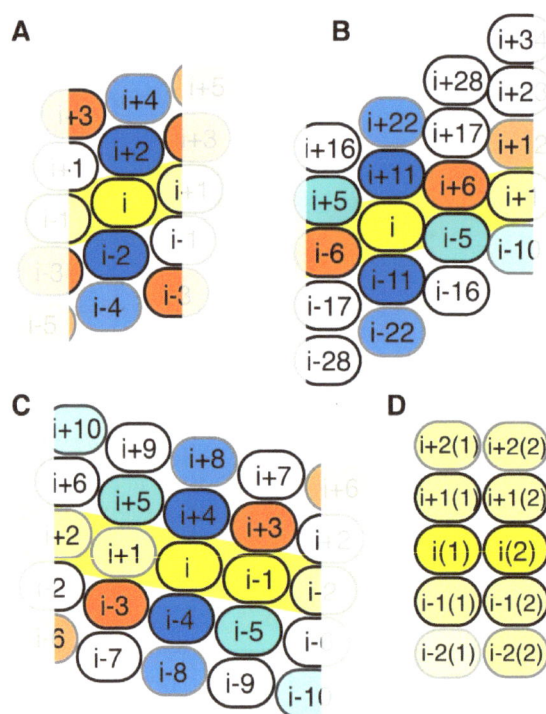

Fig. (3). Four examples of supramolecular symmetries established by ssNMR methods. (**A**) 1-start helical arrangement (such as in [41]), (**B**) 11-start helical arrangement (such as in [40, 43], (**C**) 1-start helical assembly (similar to [42]) and (**D**) ring or stacking assemblies (such as in amyloid fibrils, see for example [44]). Rise per subunit is highlighted in yellow.

In terms of other NMR-based technologies, another important upcoming system, promising also for biological material, is dynamic nuclear polarisation (DNP) ssNMR, which should be mentioned even though not further discussed in the chapter. DNP relies on the application of microwave radiation, *in situ* within the NMR spectrometer, to an appropriately paramagnetically-doped sample. This can significantly increase sensitivity and is starting to provide atomic-resolution structural information for biological samples [48 - 51].

The objective of this work primarily focuses on the investigations of self-assembling protein systems, however, many of the below-described methods will also hold true for other biological samples including sedimented monomeric proteins [52] or those in complex with membrane mimetics [11, 12, 53 - 55] such as bicelles [56, 57], glass-plate samples [58], liposomes [59, 60], nano-discs/macro-discs [61 - 63], samples in native membrane extracts [64] or even in cellular environments [6, 65].

SOLID-STATE NMR SAMPLE PREPARATION

Protein Expression

NMR spectroscopy is a mass (or in fact number of "spins") sensitive technique, wherein more sample typically yields more signal and requires less experiment time. The required mass commonly falls within the order of 0.5 to 20 milligrams for ssNMR protein samples. Therefore, the rigorous optimisation of expression conditions is an important prerequisite for NMR structural studies in order to obtain sufficient material. We will not describe the optimisation steps in detail, but by including a brief summary thereof, we intend to make the reader aware of the elements that can concertedly contribute towards improved protein yields.

Firstly, the choice of a host organism for protein expression is worth reflecting upon. Recombinant protein expression for NMR is traditionally carried out in *E. coli* as it tends to offer the best protein yield per cost in short time frames [66]. The benefits of thoroughly documented production protocols, well characterised metabolic pathways, and robust growth make it the most practical approach for obtaining high yields with isotopic enrichment. However, in cases of post-translational modifications, eukaryotic expression systems may be necessary (*i.e.* yeast [67], insect or mammalian cells [68, 69]), although they typically incur more cost, lower yields, and require stricter control of growth conditions. In cases of problematic *in vivo* expression, for example where the expressed protein is toxic to the cell, eukaryotic [70] and prokaryotic [71] cell-free expression systems are also an option. However, although these can allow greater flexibility for isotopic labelling, the start up cost and lower yields can be prohibitive.

Secondly, the DNA vector itself should be chosen with care, keeping in mind the suitability of any antibiotic resistance, the plasmid copy number and positioning of the Shine-Dalgarno sequence (in particular for *in vitro* methods), and the choice of a transcriptional promoter relevant to the host organism with desirable translational control (*i.e.* constitutive *vs.* regulated). For standard recombinant bacterial expression in *E. coli*, many commercial pDNA vectors are available especially optimised for expression, most of which employ a T7 promoter under the control of a *lac* operon for IPTG (Isopropyl β-D-1-thiogalactopyranoside) induction. The placement (N- *vs.* C-terminal) of any affinity purification tags should also be well considered, or perhaps even tested, to ensure correct native folding, protein-protein interactions, and quaternary assembly can occur unhindered. Finally, for bacterial expression specifically, the gene insert itself can also be optimised for the use of bacteria-preferred codons to achieve maximal yields. For a review of recombinant protein expression in *E. coli*, plasmid optimisation and potential troubleshooting [72].

Bacterial cultures for unlabelled protein production typically employ the use of lysogeny-broth growth medium (LB), or close variants thereof such as 2YT or Terrific Broth. Time should again be invested in optimising the cell culture protocol in terms of the pre- and post-induction incubation times and temperatures, and the media oxygenation (*i.e.* volume/vessel ratio and baffled *vs.* non-baffled flasks). Auto-induction protocols can also be of practical use, both in terms of obtaining high cell-densities and removing the requirement for manual induction [73].

Isotopic Labelling

In order to investigate biological systems by MAS ssNMR spectroscopy it is necessary to incorporate NMR-active carbon (^{13}C) and/or nitrogen (^{15}N) isotopes into the sample. This requires the substitution of the naturally abundant metabolic sources within the growth medium for isotopically enriched alternatives and, for *in-vivo* expression, is achieved through the use of Minimal Essential Medium (MEM). The standard MEM mix only includes the obligatory metal salts necessary for cell growth. Isotopically enriched metabolic sources of nitrogen and carbon (typically ^{15}N-ammonium chloride and ^{13}C-glucose) can then be included (Table **1**).

The ^{13}C- and ^{15}N-labelled sources can be incorporated individually, mixed, or with specific atom-position labelling to allow for different labelling schemes that permit different spectral correlations to reveal specific structural information (see section: **Atomic Structure Investigations**). Importantly, the optimised conditions for unlabelled cultures might not directly equate to ideal MEM conditions and some degree of reoptimisation is typically required. This is especially important as isotopically enriched components are expensive, therefore optimising the conditions for MEM expression should be carried out with naturally abundant versions of the metabolic sources. In addition, if samples are to be deuterated, appropriate D_2O adaptation must also be carried out for optimal yields [74].

Table 1. The composition of standard MEM medium for ^{13}C-^{15}N labelling.

Component	Concentration
NaCl	0.5 g.L^{-1}
KH$_2$PO$_4$	3 g.L^{-1}
Na$_2$HPO$_4$	6.7 g.L^{-1}
MgSO$_4$	1 mM
ZnCl$_2$	10 µM
FeCl$_3$	1 µM
CaCl$_2$	100 µM
MEM Vitamin mix (100x)	10 mL.L^{-1}
^{13}C-glucose	2 g.L^{-1}
$^{15}NH_4Cl$	1 g.L^{-1}

The most informative initial isotope-labelling scheme is uniform ^{13}C and ^{15}N incorporation by the addition of ^{13}C-glucose and ^{15}N-ammonium chloride to the growth medium. Experiments can then be readily acquired which correlate the chemical shifts of carbon-carbon or carbon-nitrogen nuclei throughout the protein backbone and side-chains. So-called fingerprint ^{13}C-^{13}C experiments are often sufficient to assess the structural homogeneity of the protein, provide some level of residue-specific peak assignment and give indications of the secondary structure propensity of the protein subunit in the assembly. Dependent on several parameters (such as protein subunit size, structural homogeneity and spectrometer field), the ^{13}C-^{13}C fingerprint spectra can be deleteriously overlapped. In these instances, the acquisition of spectra in both ^{13}C and ^{15}N dimensions can be advantageous. This is achieved by acquiring two or three-dimensional ssNMR experiments that correlate one (or more) carbon atoms to inter- or intra-residue ^{15}N backbone nuclei. The ^{15}N chemical shift correlations then act to discriminate between different carbon-carbon correlations (see subsection: *Secondary Structure Determination – Solid-state NMR Resonance Assignment*).

For structural analysis of protein assemblies, a crucial step is the collection and assignment of distance restraints, *i.e.* identifying short inter-nuclear proximities that define the 3D fold of the protein subunits and their placement with respect to each other. In ssNMR, internuclear proximities are commonly extracted from spectra encoding for carbon-carbon distances. However, spectra that are recorded to observe long-range ^{13}C-^{13}C correlations are often difficult to analyse due to spectral crowding. In addition, for macromolecular assemblies, it is essential to discriminate between intra- and inter-molecular distance restraints. In such cases, specific labelling schemes are very useful in order to reduce the spectral complexity. This is achieved by the use of specifically ^{13}C-labelled metabolic precursors such as [1,3-^{13}C]- or [2-^{13}C]-Gly or [1-^{13}C] or [2-^{13}C]-Glc, which serve to dilute the labelled carbon nuclei with respect to the respective metabolic pathways. Specifically labelled glycerol has the advantage over glucose that one or two out of 3 *versus* one out of 6 carbons are ^{13}C labelled per residue, leading to a better absolute sensitivity per protein subunit. This is helpful when the signal-to-noise ratio is an issue towards determining long-range restraints.

Discriminating between intra- and intermolecular restraints is an important task. If preliminary structural information is available, for example on the subunit structure or the symmetry parameters of the assembly, these can be used to disambiguate the detected long-range restraints (see subsection *Long-range Distance Restraint Collection – Intra- and Intermolecular Distances*). ssNMR approaches to disambiguate intra- and intermolecular contacts include the preparation of a (50/50) [^{13}C]/[^{15}N]-labelled sample to directly detect intermolecular restraints (Fig. **2B**) or preparing diluted (50/50) [^{13}C]-labelled / unlabelled samples, as outlined in Fig. (**2D**), which can be used to emphasise intra-molecular versus inter-molecular contacts. Likewise, preparing mixed (50/50) [1-^{13}C]-Glc and [2-^{13}C]-Glc labelled samples for the acquisition of intermolecular restraints can be of use (Fig. **2C**) [34].

Proton detection is becoming increasingly accessible by ssNMR from methodological improvements in probe design, which allow faster MAS speeds to be obtained ($\geq$ 60 kHz) that diminish the strong [1]H-^{1}H dipolar couplings. However, proton detection in MAS regimes below 60 kHz is still possible and can yield well-resolved spectra when ^{1}H nuclei are significantly eliminated from both the sample and solvent by substitution with ^{2}H deuterons [21, 75, 76]. For the complete removal of protons (perdeuteration), well-described protocols exist for bacterial protein expression [74, 77]. In these cases, small volume cell cultures in rich and minimal media (15 ml per culture) are used to acclimatise the bacterial cells to increasing proportions of [2]H_2O deuterium oxide. For example, one unlabelled rich medium preculture can be used to inoculate a protonated MEM culture, which is then taken to inoculate a penultimate 50% [2]H_2 MEM culture that is then finally harvested by centrifugation and resuspended in the desired [2]H [13]C- and/or ^{15}N-labelled MEM media, with [2]H-labelled carbon sources. Protons to be studied are then reintroduced by proton "back-exchange", *i.e.* by resuspending the purified sample or by purifying, to avoid a lack of protons in the protein core, in a [1]H_2O-containing buffer to allow labile HN moieties to exchange and achieve a final protonation level of ~20 % [1]H [78 - 82]. In some circumstances, the use of highly restricted isotope labelling schemes may be beneficial for the identification of specific correlations, such as long-range contacts in a structure or interactions between components, or for the simplification of ssNMR spectra. Sparse labelling schemes can include the use of metabolic precursors that enable the precise incorporation of isotopes at specific functional groups. For example, the addition of both ^{13}C α-ketoisovalerate and α-ketobutyrate to bacterial culture allows for the specific incorporation of ^{13}C ILV methyl groups and has been used to study protein structure and dynamics [83].

Furthermore, it is also possible to incorporate ^{13}C and/or ^{15}N labelled amino acid types in an unlabelled background. This can be achieved by using *E. coli* strains auxotrophic for individual or various combinations of amino acid types which, when added in their labelled forms to unlabelled MEM cultures, are then incorporated into translated proteins [84]. This type of amino acid specific labelling can also be accomplished with cell-free expression systems where labelled amino acids can be simply added to the cell-free mix in lieu of the naturally abundant alternatives [85 - 87].

Purification

After fully optimising the expression of a recombinant protein, combined with the costs associated with isotope labelling, purifying the maximal quantity of a protein, in its purest form, is paramount. For the standard purification of soluble or stable monomeric proteins, wide ranging combinations of chromatographic (*e.g.* affinity, ion-exchange, hydrophobic interaction and size-exclusion chromatography (SEC)) and biochemical (precipitation, centrifugation) protocols exist and are well documented. The same methods also apply to the purification of self-assembling protein subunits, if soluble, and therefore won't be discussed in detail. However, for systems that assemble *in vivo*, it is sometimes necessary to modify the conditions or to add steps to optimise the results.

Firstly, such systems are often found to accumulate inside bacterial inclusion bodies and therefore require additional lysis steps to fully liberate the expressed proteins [88]. The internalisation of proteins within inclusion bodies can be investigated by SDS-PAGE. If necessary, the resuspension of cell-lysis pellets in strong denaturing conditions (*e.g.* 8 M guanidinium HCl, 150 mM NaCl) is adequate for protein extraction. Secondly, it is essential to inhibit self-assembly for chromatographic purification, to keep purification tags accessible, maintain sample homogeneity and allow for the unhindered flow of sample and solvent within the apparatus. Denaturing conditions during purification should ensure this, if applicable, and any purification apparatus should be equilibrated and run taking the denaturing buffer conditions into account. Purification should be followed by analysis of the sample purity (for example by SDS-PAGE), however, as protein self-assembly can also act as a final purification step, purity should also be checked after assembly of the final objects (see below).

Protein Self-assembly

Following purification, buffer or solvent exchange is usually required for protein assembly to occur. This can be achieved by dialysis immediately after purification to remove any remaining denaturants and/or to adjust the pH. There cannot be an all encompassing methodology to predict suitable protein assembly conditions [89], indeed some conditions may promote non-native or polymorphic assembly. Therefore, finding optimal conditions for the assembly of novel systems usually involves screening different pH, ionic strength, and/or reducing conditions with unlabelled samples (if doubts remain, labelled) and, if assembly occurs, samples can be tested by one-dimensional ssNMR (if labelled, multidimensional ssNMR). After adjustment of the aqueous conditions, proteins are usually concentrated and set aside at room temperature in small volumes for 1-10 days to assemble. Agitation during assembly has been shown to increase the assembly rate of some systems, in particular for amyloid fibers [90], and optimisation of incubation temperature can also prove worthwhile.

Typically, protein self-assembly is accompanied by the aqueous sample becoming opaque due to the formation of larger objects. The rate of assembly may be monitored during incubation by solution NMR [91] or, for example with amyloids, from the change in intrinsic fluorescence associated with aggregation [92]. Once assembly has been achieved, samples are harvested by ultracentrifugation. If possible, sample homogeneity and quality should be verified with electron-microscopy (EM) and ssNMR (see section: **Global Architecture Assessment**).

In reference to the previous section, protein self-assembly can also provide some degree of sample purification. The sedimentation of a molecular assembly during centrifugation often leaves any soluble impurities in the supernatant. In fact, if the sample can tolerate multiple rounds of assembly and disassembly, elimination of contaminants can be achieved from relatively impure samples. The purification of actin described by Pardee & Spudich [93] is one such example where, from natural sources, the repeated assembly and disassembly of actin filaments is sufficient to drive protein purification, both from other assorted muscle proteins and biologically inactive actin monomers. Nonetheless, the purity of the final assembled object should always be checked by SDS-PAGE.

ssNMR Rotor Filling

The final stage of sample preparation concerns the packing of the harvested sample into the "rotor" for ssNMR experimentation. Rotors are cylindrical vessels that are closed with a cap decorated with a series of wings that allow the sample to be precisely spun at the magic angle inside the magnet from an incoming stream of air within the probe. Typical rotor diameters for biological samples range from 0.7 mm to 4 mm (Table **2**) and can accommodate various sample volumes (0.5 - ~70 µl). Before filling, samples should be ultra-centrifuged to achieve highest sample compactness. Common additions to samples ready for packing include 0.02% (w/v) sodium azide (NaN_3) to avoid microbial growth and DSS (4,4-dimethyl-4-silapentane-1-sulfonic acid) as an internal chemical-shift reference [94]. Likewise, rotors should be thoroughly cleaned before use (even when new) to help eliminate the detection of impurities. This is especially true for ^{1}H-detection, owing to its enhanced sensitivity, and standard practice should be to wear gloves when handling the rotor to reduce the possibility of contaminants. Samples are added to rotors either manually, typically with a spatula and repeated rounds of bench-top centrifugation to eliminate the supernatant, or using specialised ultra-centrifuge units designed to pack near-solid or gel-like samples [94 - 96]. This is especially necessary in the cases of small rotor sizes (0.7 – 1.3 mm diameter). The choice of rotor size depends on the available ssNMR probe hardware, as these are precisely matched to specific rotor sizes; the smaller the rotor diameter, the higher the possible MAS frequency. With the smallest available rotors (0.7 mm diameter), ultra-fast spinning speeds of ~111 kHz can be achieved for direct proton detection and increased experimental sensitivity. Higher spinning frequencies can also practically accommodate reduced sample amounts in terms of signal-to-noise due to the improvements in peak linewidths.

Table 2. ssNMR rotor sizes, volumes and maximum MAS speeds. Stated sample volumes are approximate. *Thin walled rotors (> volume) or the use of internal spacers (< volume) can be used to adapt volumes.

Rotor Diameter	Total Volume*	Maximum MAS
4 mm	~ 70 µl	~14 kHz
3.2 mm	~30 µl	~22 kHz
2.5 mm	~10 µl	~35 kHz
1.3 mm	~ 2.5 µl	~65 kHz
0.7 mm	~0.5 µl	~110 kHz

GLOBAL ARCHITECTURE ASSESSMENT

From the simplest 1D ssNMR experiments the observable features of chemical shifts and line-width are highly informative probes for global molecular structure and dynamics [97]. Local structural information such as secondary and tertiary structure, as well as the identification of quaternary assembly or interactions with accessory molecules first requires chemical shift assignment and then restraint or interaction detection and interpretation. In most cases these tasks require the acquisition of multidimensional ssNMR spectra, their analysis and/or selective labelling strategies. However, the first structure assessment can be performed on unlabelled samples of the protein assembly and provides an efficient and cost-effective method to assess sample purity, homogeneity and structural order (as described in subsection *Protein Self-Assembly*). Therefore, two types of 1D ssNMR experiments are set up and acquired (for a protocol see [20]), a ^{1}H-^{13}C cross-polarisation (CP) to probe rigid segments in the assembly architecture and a ^{1}H-^{13}C INEPT to reveal mobile segments that are strongly reduced in CP-based experiments.

Two example 1D ^{1}H-^{13}C CP spectra of amyloid fibrillar assemblies are presented in Fig. (**4**) highlighting the regions of protein Cα and aliphatic carbons on the carbon chemical shift scale. The 1D ^{13}C spectrum in Fig. (**4A**), left, reflects a first fibril preparation of the HELLP functional amyloid protein domain (~8 kDa) from *Chaetomium globosum* and shows an intermediate spectral resolution [98]. Nevertheless, the high signal intensity and the here-observed spectral resolution for a protein assembly of this molecular size indicate a structured architecture and shows promise as a candidate for multidimensional ssNMR structure analysis. In this case, 2D ssNMR (^{13}C-^{13}C correlation experiments) can prove useful, for instance in characterisation of the secondary structure elements, as was indeed the case for HELLP [98]. However, in high-resolution ssNMR pursuits such as structure determination, the requirement for thorough resonance assignment would necessitate the modification of the construct and/or assembly protocol. In comparison, the 1D ^{13}C spectrum of the HET-s functional amyloid prion from *Podospora Anserina* (Fig. **4A**, right) shows very sharp NMR signals for the same ^{13}C aliphatic and Cα spectral region. The inference for this example would be that the protein subunits in the fibrils are highly ordered and homogenous, and that additional multidimensional spectra would be suitable to resonance assignments and subsequent atomic-level structural studies. Indeed, HET-s fibrils have been extensively studied, its structure has been determined by ssNMR [99], and it is considered as a reference in terms of ssNMR spectral quality.

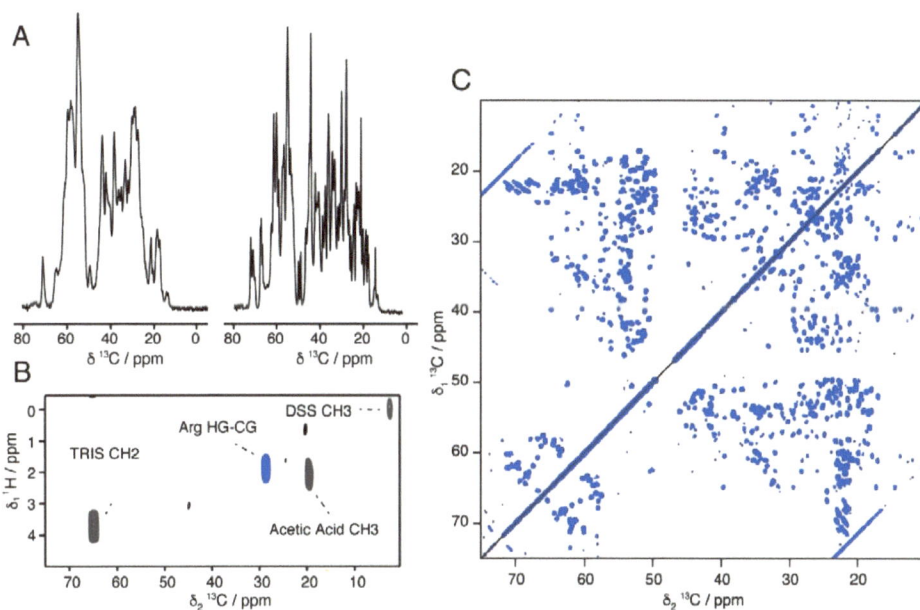

Fig. (4). Examples of 1D and 2D spectra for the global assessment of protein structure. (**A**) 1D ^{13}C {H}C CP experiments for the HELLP protein (left) versus the fungal HETs functional amyloid (right). (**B**) 2D {H}CH INEPT experimentation for a sample of the HELLP protein fibrils. (**C**) A 2D CC PDSD spectrum (50 ms mixing time) for the type 1 pilus from *E.coli* [41].

The acquisition of INEPT based 1D or 2D ^{13}C spectra (Fig. **4B**) is then informative in characterising the mobile regions (with respect to isotropic motions) of the sample. In Fig. (**4B**), this is highlighted for the HELLP protein fibrils [98] by the observation of CH correlation peaks corresponding to buffer components and Cγ-Hγ arginine side chains. No major mobile regions can be detected in this protein assembly.

If the preliminary results are promising towards more detailed structural investigations, as is the case for the two here-selected examples (Fig. **4A** and **B**), the production of uniformly ^{13}C,^{15}N, sample is the next step for the assessment of the assembly structure. The acquisition of ^{13}C-^{13}C correlation experiments, a spectral fingerprint of the sample, allow for global structure characterisation and primary evaluations of local-level secondary structure content. An example is shown here for the ^{13}C-^{13}C PDSD experiment (proton-driven spin-diffusion [100]) of the type 1 pilus from *Escherichia coli*, with its protein subunit FimA (Fig. **4C**) [41].

The 2D ^{13}C-^{13}C PDSD experiment detects all proximal ^{13}C nuclear correlations, through-space, and by restricting the carbon-carbon mixing time to short values (here 50 ms), we focus on the appearance of intra-residue ^{13}C-^{13}C correlations. Inter-residue correlations to neighbouring residues can still occur, but they are usually few in number and very low in intensity compared to the intra-residual signals.

From a cursory analysis of the spectral quality, the evident characteristics of the sample represented in Fig. (**4C**) are that:

1) the large number of visible signals with respect to the protein subunit size (FimA, 16 kDa) shows that a large portion of the protein in the sample is in a rigid conformation; 2) signal linewidths with respect to the acquisition parameters (spectrometer field, MAS frequency, decoupling strength) indicate the protein subunits adopt very homogenous structures; 3) spectral dispersion indicates a structure rich in β-strand conformation, *i.e.* a large number of signals are observed in spectral regions typical for β-strand conformations, although a few signals also reflect α-helical or random-coil conformations.

Conclusions based on the number of observed peaks, their line-widths, and their good dispersion, indicate that a detailed analysis and atomic-specific resonance assignment should be possible.

From this type of well-resolved spectrum, it is possible assess the visible amino acid types owing to their distinct chemical shift values and signal correlation patterns and their numbers. With knowledge of the primary sequence, conclusions can then be drawn as to extent of polymorphism within the sample, what proportion of the sample is forming the rigid components of the assembly and, globally, which sequence-specific regions could be involved in the rigid assembly core. For example, polymorphism in the protein structure of the assembly exists if the number of observed amino acid specific signals is greater than the number of those amino acids in the primary sequence. If, on the other hand, fewer peaks than expected are observed for a given amino acid type or many peaks are not assignable due to low signal intensity, dynamics can be the at the basis of these phenomena and reducing the experimental temperature can serve to dampen backbone dynamics and improve resonance signal observation in solid samples. Furthermore, from the identification of amino acid type and Cα/Cβ chemical shifts, the extent of helical or strand secondary structure elements can be estimated from determining the secondary chemical shift. This can be done by manual comparison to known helix or strand-specific reference values [101] or by automated tools such as the CSI webserver [102]. In instances of unique amino acids in the primary sequence, the corresponding signals can be identified unambiguously. It may occur that sequential residue pairs can be identified from short mixing time PDSD spectra through the appearance of the weaker inter-residue correlation peaks.

Without the acquisition of additional 2D spectra (*i.e.* longer mixing time PDSD) or the production of samples with alternate labelling-schemes, it can be difficult to make further, higher-resolution, conclusions at this stage. However, the use of *in silico* tools can provide an informative means in a time and cost effective manner.

If structures were determined by NMR methods, for example of soluble homologues, or mutated or truncated versions of the assembly subunit protein, access to deposited chemical shifts is straightforward from the BMRB databank [103] and NMR software packages such as CCPN Analysis [104] allow for their straightforward input and overlay onto ssNMR data for comparison. In cases where homologous structures have been determined by a different method, several webservers such as SHIFTX2 [105] can back-calculate chemical shifts from any pdb structure. However, the predicted shifts should be considered with caution because chemical shifts of complex systems cannot, to date, be predicted accurately.

Another powerful approach is the calculation of position-specific secondary structure probabilities from the protein amino-acid sequence alone with online tools such as PSIPRED [106]. These are certainly not fool-proof, however they can help to estimate the amount and position of rigid secondary structure elements that could be observed in the spectra. It is then also possible to back-calculate average chemical shifts for the expected ordered residues and see how they compare to the data.

ATOMIC STRUCTURE INVESTIGATIONS

Secondary Structure Determination – Solid-state NMR Resonance Assignment

The next level of structure determination by ssNMR allows for the residue-specific determination of secondary structure from the analysis of assigned chemical shifts. As previously described, global structural features of the protein assembly can be identified in a fingerprint ^{13}C-^{13}C correlation spectrum which utilises C-C mixing schemes such as PDSD, DREAM (dipolar recoupling enhancement through amplitude modulation [107, 108]) or DARR (dipolar-assisted rotational resonance [109]). However, in order to approach the sequential resonance assignment, additional experiments must be acquired to allow the specific identification and assignment of intra- and inter-residue peaks. Firstly, similar versions of the 2D CC PDSD spectra can be acquired, with extended CC mixing times. At intermediate mixing times (usually 100-200 ms), more distant, inter-residue, C-C correlations can be observed, which become clearly evident when compared to short-mixing time experiments.

Fig. (**5A**) shows one example of a sequential resonance assignment using two PDSD spectra with distinct mixing times for the type 1 pilus from *E.coli,* where both the Cα and Cβ nuclei of a serine residue show correlations to an adjacent glycine Cα nucleus. For large systems, peak crowding may make the study of 1,3-^{13}C-Gly or 2-^{13}C-Gly spin-diluted samples necessary for the sequential resonance assignment using 2D CC correlation experiments as well as for the unambiguous assignment of side-chain carbons.

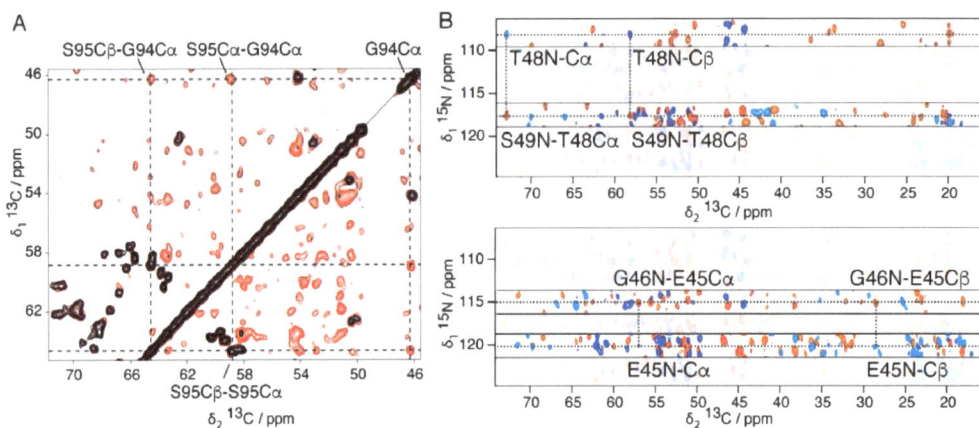

Fig. (5). Example 2D assignment strategies from a uniformly [^{13}C,^{15}N] labelled sample of the type 1 pilus from *E.coli.* (A) A short range (50 ms mixing time) 2D PDSD experiment (black) overlaid with a long range (175 ms mixing-time) PDSD experiment (red) showing sequential Cα-Cα and Cβ-Cα correlations for S95 and G94. (B) 2D NC assignment strips highlighting inter-residue NC correlations between T48-S49 (top) and E45-G46 (bottom) for NCACB (+blue/-light blue) and N(CO)CACB (+red/-orange) experiments.

With increasing size of the protein subunits and dependent on the spectral quality obtained for the assembly, the spectra become more complex and the sequential resonance assignment requires additional data, as was the case for the type 1 pilus with its subunit FimA (16 kDa). The use of 2D heteronuclear experiments, involving ^{15}N-^{13}C polarisation transfers, is a complementary approach employed during resonance assignment. Fig. (**5B**) shows a typical sequential assignment using NCACB and N(CO)CACB experiments [110] of the FimA pilus, which correlate intra-residue (*i*) and preceding-residue (*i-1*) carbon atoms to a given residue's amide nitrogen, respectively. Proceeding from the NCACB (N, Cα and Cβ of residue i) to the N(CO)CACB (N of the residue i+1, Cα and Cβ of residue *i*) and back to the NCACB (N, Cα and Cβ of residue *i+1*), allows consecutively assigning the visible resonances to the corresponding atoms on the primary protein sequence.

In conjunction with 2D CC correlation spectra (short and long mixing times), the information on Cα and Cβ of one residue in the NC spectra can be combined with the side-chain carbons that might show correlations with the neighbouring residues in the CC spectra. They also provide a powerful method to identify spin systems and facilitate their assignment.

When spectral crowding remains problematic, three-dimensional variants of the NCACB and N(CO)CACB (or N(CO)CX) experiments can be recorded and supplemented by a wide array of complementary heteronuclear experiments which can disambiguate assignments between spin systems that share similar resonance frequencies during the sequential assignment process (Fig. **6**). To obtain unambiguous sequential assignment, it is usually required that two resonance frequencies remain constant whereas one varies in one assignment step.

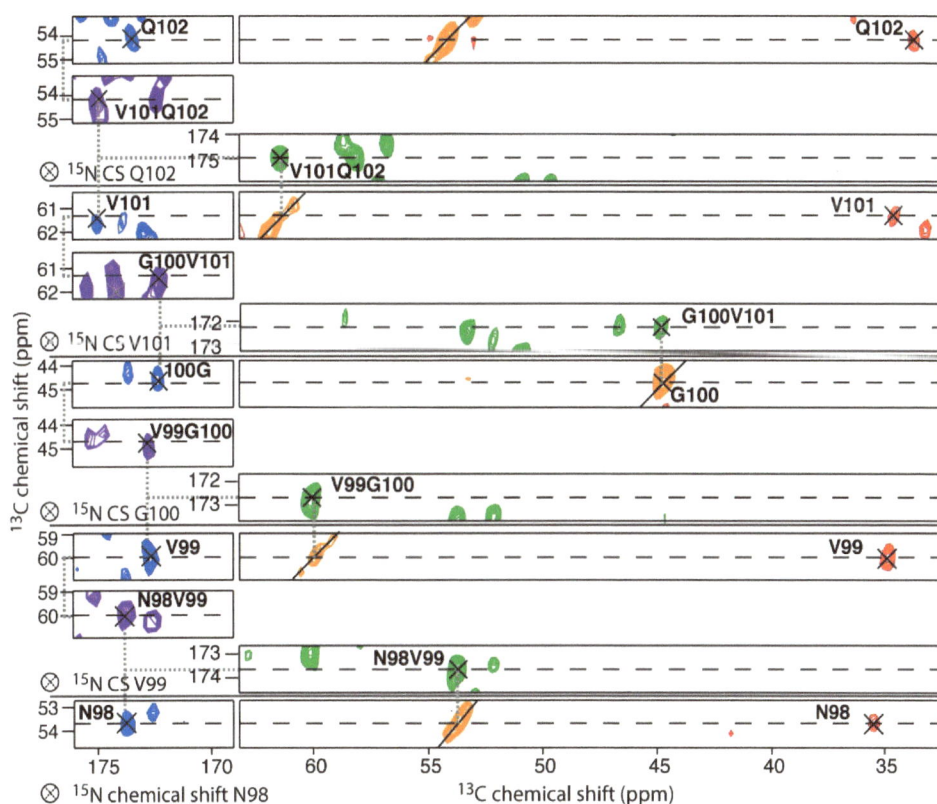

Fig. (6). Example 3D assignment strategies from a uniformly [^{13}C,^{15}N] labelled type 1 pilus from *E.coli*. Five groups of 2D strips are shown, at the five corresponding ^{15}N amide resonant frequencies for Q102 through N98 (top to bottom). Spectra are colour coded as follows: NCACB (+orange/-red), NCACO (+blue), CANCO (+purple) and NCOCA (+green).

Example experiments include the NCACB (intra-residue [110]), NCACO (intra-residue [110]), CANCO (inter-residue [111]) and NCOCA (inter-residue [110]), highlighted in Fig. (**6**), which offer assignment strategies akin to the liquid-state NMR experiments of proteins by their ability to navigate the different resonant frequencies of the protein backbone with logical directionality. The additional indirect spectral dimension of the three-dimensional data helps to facilitate assignment by removing ambiguities between residues with, for instance, similar amide ^{15}N frequencies. This aspect in particular can be appreciated by comparing the spectral complexity between Fig. (**5B**) (NCACB, blue/light blue) and Fig. (**6**) (NCACB, orange/red).

If the necessary hardware is available, ^{1}H-detected ssNMR experiments are very useful for resonance assignment and have been developed over the last years [28, 112 - 117]. Owing to improved sensitivity of ^{1}H detection and the inherent improvements in resolution at the high MAS speeds, these can be particularly informative for complex systems. Fig. (**7**) highlights an example of ultrafast-MAS experimentation wherein (H)NCAH and (H)N(CO)CAH spectra provide inter and inter-residue chemical shift information for the Cα-H moieties of sequential residues [28].

After thorough resonance assignment all observable peaks in the fingerprint CC spectrum (*e.g.* short mixing time PDSD) should be assigned. Unassigned peaks can create problems during the assignment of further structural restraints because any unassigned frequency creates an unassigned ambiguity. If all atoms in the supramolecular assembly were in a rigid regime, all should be assignable with the above-mentioned techniques. However, a complete assignment is not always achieved, since in particular side-chain carbons are in many cases ambiguous or in a less rigid regime and therefore not observed.

Mobile regions of the protein in the assembly can be detected in INEPT (^{1}H-^{13}C correlation) or for example INEPT-TOBSY or TOBSY (^{1}H-^{13}C or ^{13}C-^{13}C detected) [118, 119] spectra and can be amino acid specific assigned based on available random coil shifts [103] or sequentially as described in [119] or [120].

In reality, it is highly likely that some regions within a sample will be unaccounted for and that some peaks will be unassignable. During structure calculation, flexible regions can mostly be omitted since they do not establish long range contacts with the core. Importantly, biased or speculative attempts to assign resonances are likely to be incorrect and should be avoided since misassignment can create severe problems during subsequent analyses.

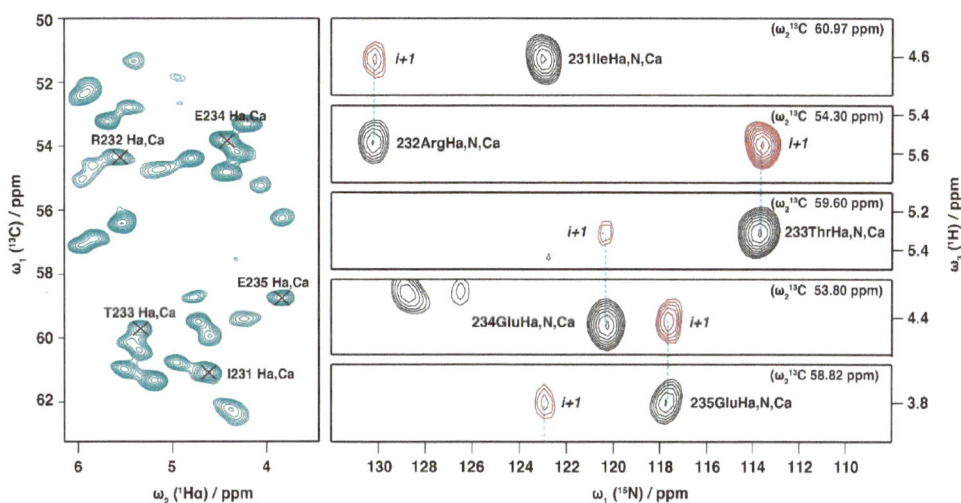

Fig. (7). An example 3D assignment strategy using proton detected experiments for the uniformly [^{13}C,^{15}N] labelled HET-s (residues 218-289) from *P. anserina*. The {H}CH correlation spectrum of HET-s (left, teal). Five 2D strips are shown (right), at the five corresponding ^{13}Cα resonant frequencies for I231 through E235 (top to bottom) with the *i* and *i+1* Hα-Cα correlations in (H)NCAH (black) and (H)N(CO)CAH (red) proton detected experiments, respectively [28]. Additional assignments have been omitted for clarity.

Following sequence specific assignment, the process of protein secondary structure characterisation relies on *in silico* evaluations. Analyses of the specific backbone and Cα/Cβ chemical shifts (either manually, or with tools like CSI) as described here-above (see section: **Global Architecture Assessment**), allows for the accurate description of secondary structure across the assigned portions of the protein. The prediction of dihedral angles can also be used to infer secondary structure when using programs such as Talos+ [35] and DANGLE [121]. The predicted dihedral angles are also included later as restraints during structure calculation (see-below, see subsection: *Distance Restraint Collection – Intra- and Intermolecular Distances*). These results can represent important first steps in the description of novel systems or as metrics for qualitative investigations of mutants or structural isoforms. Furthermore, accurate descriptions of secondary structure can be an asset for homology modelling in aspects such as the localisation of different structural elements or, conversely, whether any genetic insertions or deletions are buffered into flexible protein regions to maintain conserved core folds.

Near-complete backbone and Cα/Cβ resonance assignments of rigid structural cores can also be used for *de novo* structure calculation with certain *in silico* methods that exploit the strong relationship between structure and chemical shift. Programs such as CHESHIRE [122] and CS-Rosetta [123] function by building structural fragments from comparisons between the determined assignments and those associated to protein structures deposited online. The assembly of these fragments and their energy minimisation is then undertaken to arrive at potential structural ensembles. This type of structure calculation has been shown possible from ssNMR-determined chemical shifts in particular, for both monomeric proteins [124] and amyloid fibrils [125]. The accuracy of these structures should however be treated tentatively in the absence of any structural restraints, although methods such as these can provide important insight when more complex analyses are prohibited.

Distance Restraint Collection – Intra- and Intermolecular Distances

The calculation of high-resolution protein structures by any NMR method fundamentally involves the identification of inter-atomic distances and angles. The generation of angular restraints for protein assemblies by ssNMR is reliant upon chemical shift-based prediction tools such as Talos+ [35] and DANGLE [121], processing the backbone and $C\alpha/C\beta$ chemical shifts, to calculate backbone dihedral angles. If the angular restraints and chemical shifts imply that prominent α-helical regions exist, appropriate hydrogen bond restraints can be defined during structure calculation and other bonds or nuclear proximities, such as disulphide bonds, can be introduced where required.

The structure calculation is mainly based on long range distance restraints, *i.e.* atom contacts between residue i and residue $i \pm >4$. The collection of distance restraints is, to date, mainly based upon the examination of CC correlation spectra which encode directly detected CC distances, or indirectly detected HH distances [126]. The procedure is similar to sequential assignment wherein the user scours such spectra to identify cross-peaks, which correlate to atom contacts between residue i and residue $i \pm >4$. The use of data visualisation programs such as CCPN Analysis, optimized for solution and ssNMR studies towards protein structure and interaction studies, is advantageous in this process, especially for maximally assigned samples, wherein assignment closeness can be immediately evaluated when picking correlation peaks. Importantly, any assignment ambiguity between a long range (i-$i \pm >4$) and closer contact (i - $i \pm \leq 4$), is solved by assigning to the (i - $i \pm \leq 4$) contact. Nonetheless, depending on the protein size and the data quality, the spectra from uniformly $^{13}C,^{15}N$-labelled samples might become too complex and overlapped, and the use of spin diluted samples (see subsection: *Isotope Labelling*) has become a standard approach for collecting a maximal amount of distance restraints, especially in the investigation of large protein systems. A complementary approach is afforded by the acquisition of 3D spectra to reduce problematic spectral crowding. They can be acquired on $^{13}C,^{15}N$-labelled or spin diluted samples. Fig. (**8**) presents a simulated example of a 3D NCACX experiment acquired for the assembled PrgI Type-III secretion (T3SS) needle from *Salmonella typhimurium* [127]. The sample is spin diluted by use of 2-^{13}C-Gly labelling and the presented experimentation employed a long CC mixing time (850 ms) in order to derive specific long-range restraints. Based upon these data, the specific assignment of previously ambiguous peaks was possible and lead to the identification of novel distance restraints for the PrgI needle.

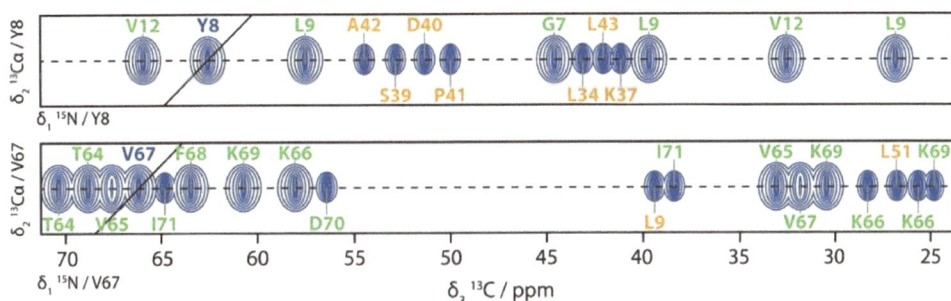

Fig. (8). A simulated example of long-range distance identification from a 3D NCACX correlation spectrum acquired with a long CC mixing step. The novel identification of long-range distances (orange labels) for the T3SS needle with such experimentation was made possible with the use of 2-^{13}C-Gly spin dilution. Figure based upon [127].

A major challenge in ssNMR-based structure investigations consists of categorising all collected distance restraints into intra- and intermolecular restraints in the supramolecular assembly. If biophysical data from different sources are available, such as a truncated or mutated protein subunit structure, the structure of a homologous system, a cryo-EM map, SAXS data, or mass-pe--length values, these can be used to help resolve ambiguous intra- or intermolecular restraints. There are also a number of ssNMR approaches that can be employed to identify inter- and intra-molecular restraints specifically.

As outlined in Fig. (**2B**), preparing diluted 50:50 ^{13}C-labelled / unlabelled samples can be used to emphasise intra-molecular contacts by reducing the populations of adjacent labelled monomers. Likewise, preparing 50:50 mixed 1-^{13}C-Glc and 2-^{13}C-Glc-labelled or 1,3-^{13}C-Gly and 2-^{13}C-Gly samples (Fig. **2C**) for the acquisition of long mixing time spectra aids the observation of intermolecular restraints. It must be stressed however that care must be paid to check for consensus between datasets and between different labelling schemes before defining restraints. Similarly, inconsistencies should be thoroughly investigated and, if necessary, the backbone assignment re-evaluated. Restraint collection can be seen as an iterative process including all available information during assignment.

During the distance restraint assignment process, there usually remains some degree of resonance assignment ambiguity. If these cannot be resolved by logical inferences from all available data and information, ambiguous restraints can be assigned as such and re-evaluated during the structure calculation process itself.

More recently, the advent of ultra-fast MAS proton-detected ssNMR has allowed for high-resolution spectra to be acquired which, owing to the superior sensitivity of ^{1}H and their greater ubiquity can give rise to otherwise unobtainable information in practical acquisition times. Fig. (**9**) presents a simulated example of an RFDR-based 3D (H)CCH total correlation spectroscopy (TOCSY) spectrum acquired for the *Acinetobacter* phage coat protein AP205CP [27]. The sample was uniformly ^{13}C,^{15}N-labelled, fully protonated and, at the applied 100 kHz MAS rate, allowed for direct ^{1}H detection and the observation of highly resolved peaks. Pintacuda *et al.* were able to combine distance restraints from multiple data sets acquired with ultra-fast MAS (H)NHH, (H)CHH and H(H)CH experiments, along with predicted dihedral angles, to generate well performing structural ensembles for both the AP205CP protein and the model protein GB1 without the need for ^{13}C-detected experimentation. Impressively, the relative alignment of the dimeric AP205CP subunits was also identified from these data and, in doing so with a single ^{13}C,^{15}N-labelled sample, represents an important emerging ssNMR strategy if the hardware is available.

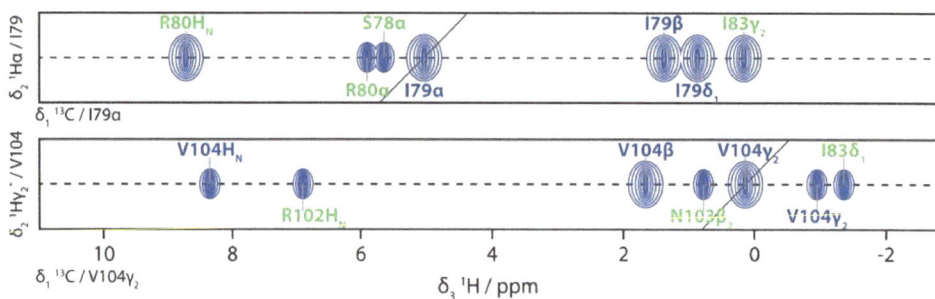

Fig. (9). A simulated example of long-range distance identification from a proton detected RFDR (H)CHH TOCSY spectrum (1 ms mixing time) of the AP205CP coat protein, uniformly ^{13}C,^{15}N-labelled, from *Acinetobacter* phage AP205. The identification of intra- (blue labels) and inter-residue (green labels) proton distance restraints and calculation of dihedral angles can alone be sufficient for structure determination by ssNMR. Figure based upon [27].

Structure Modelling

The geometric restraints obtained by ssNMR methods are used in conjunction with computational simulated annealing protocols to effectively fold protein(s) *in silico* in order to arrive at structures that are both realistic in relation to known chemical bonding and geometry, and representative of the experimental data.

Comprehensive molecular dynamics methodologies [38, 39, 128] and software packages [37, 129, 130] now exist which incorporate energy terms to account for a wide array of NMR-determined structural parameters, as well as those from other biophysical techniques such as cryo-electron microscopy, solution scattering or liquid state NMR [131 - 133].

Long-range distance restraints are the primary driving force for protein structure determination by ssNMR. In practice, restraints take the form of a formatted text file which, for each restraint, includes a reference to two atomic sites (both positions can be specific or ambiguous), and either upper and lower distance boundaries in Ångström (*i.e.* Cyana formatting) or a primary distance with separate upper and lower distance bounds (*i.e.* CNS/XPLOR formatting). File formats for other restraint types (dihedral angles, hydrogen bonds, *etc.*) can be found online. In determining the specific distances to use for restraints, it is possible to calibrate ssNMR cross-peak intensities to distances [134] based upon the fundamental $1/r^6$ relationship which governs the strength of the internuclear dipolar interaction. However, in reality, unlike liquid-state NMR experimentation where this approximation can be calibrated for individual datasets, standardisation is still unreliable in ssNMR conditions due to differences in sample dynamics and magnetisation transfer efficiencies. Therefore, for any given ssNMR distance restraint, the prevailing methodology is to initially define all restraints within 0 to 8 or 10 Å. Lower distance bounds can be treated as the Van der Waals radius (*i.e.* 1.8 Å for ^{1}H-^{1}H distances), although removing lower-bound restrictions can allow for better diastereotopic modelling.

The computational modelling of the protein or assembly is then carried out as a series of iterative steps, with respect to the minimisation of energy functions for the chemical and experimental parameters and the optimisation of the ambiguous restraints. A number of ambiguous assignment possibilities for a given restraint can usually be discarded due to the significance of other restraints. However, if multiple restraint ambiguities can be accommodated, closer inspection of the ssNMR data should again be carried out manually. Likewise, when a particular restraint cannot be satisfied at all or is significantly violated as a result of the fold described by the remaining data, incorrect assignment of the restraint or perhaps even the initial assignment is often the case and data should again be manually evaluated.

The computational speeds now available in standard desktop machines and the cluster-based computing commonly offered in academic institutions has made protein structure determination possible within a matter of hours. Nonetheless, when facing the iterative optimisation of input data in regard to output structures, multiple calculations can prove time consuming. For this reason most computational methodologies model protein structure at two fundamental levels. The most simplistic form is torsion-angle dynamics (TAD) and aims to satisfy the available restraints by considering the protein as a simplified series of interconnected nodes, usually *in vacuo*. The second and more rigorous form, full-atom Cartesian dynamics, models the full atomic structure of the protein and includes specific force-fields for inter-atomic forces. Although TAD is comparatively less precise, this is at the expense of less computational time and can therefore provide feedback on restraint ambiguities and optimise inter-nuclear distances in practical timeframes. Subsequent modelling can then be carried out in Cartesian space with solvent effects also taken into consideration to refine the initial TAD structure.

Typical structure calculation protocols with NMR data calculate >100 final structures, independently initiated from randomised coordinates, which are then described or deposited online as a group (or *ensemble*) of ~10-20 structures. Judging which structures, or conformers, should be included in the final ensemble can achieved by ranking the best energetically performing structures or those with least restraint violations. The purpose of presenting an ensemble as compared to a single best performing structure is to highlight any structural heterogeneity that can be accommodated by the experimentally determined restraints or conversely, that is prescribed by the scarcity of restraints. Careful scrutiny of the modelled conformers versus the experimental data is essential in order to describe regions that are heterogeneous compared to regions imprecisely or insufficiently defined.

Finally, after the structure calculation process, the biophysical quality of the resulting ensemble should be validated by tools such as the PSVS [135], WHATIF [136], and MolProbity [137], which combine multiple structural analyses of the ensemble to judge the structural quality. These tools can highlight problems such as unrealistic dihedral angles (Ramachandran analyses), inconsistencies with original experimental data or general structural irregularities as compared to high-resolution crystallographic structures. Minor violations or steric clashes can indicate problematic restraints (*i.e.* too stringent distance bounds) or perhaps incorrect assignment of the restraint or indeed even the root resonances. Structure validation should be considered yet another element in the iterative cycle of data refinement for optimal structure calculation.

Three such examples of modelling with ssNMR data are included below (Fig. **10**). Fig. (**10A**) represents the T3SS needle from *Salmonella typhimurium* [40]. In this study, a combination of ssNMR, electron microscopy and *in silico* modelling was used to provide near-atomic level information. ssNMR resonance assignment and distance restraint identification for the system was made possible by a series of ^{13}C spin-diluted samples, from which >500 combined inter- and intramolecular short and long range restraints were obtained. Resonance assignment yielded the secondary structure of the individual PrgI needle subunit and was used to identify position-specific structural fragments with the Rosetta fold-and-dock protocol [138]. The protocol then incorporated the ssNMR restraints and trialled a number of modelling runs to identify both the most favourable handedness and helical symmetry of the needle in agreement with the NMR data and scanning transmission electron microscopy (STEM) measurements of the axial rise per subunit. The radius of the modelled 29-subunit needle was restrained during the calculation in regards to a published cryo-EM map of the assembly [139].

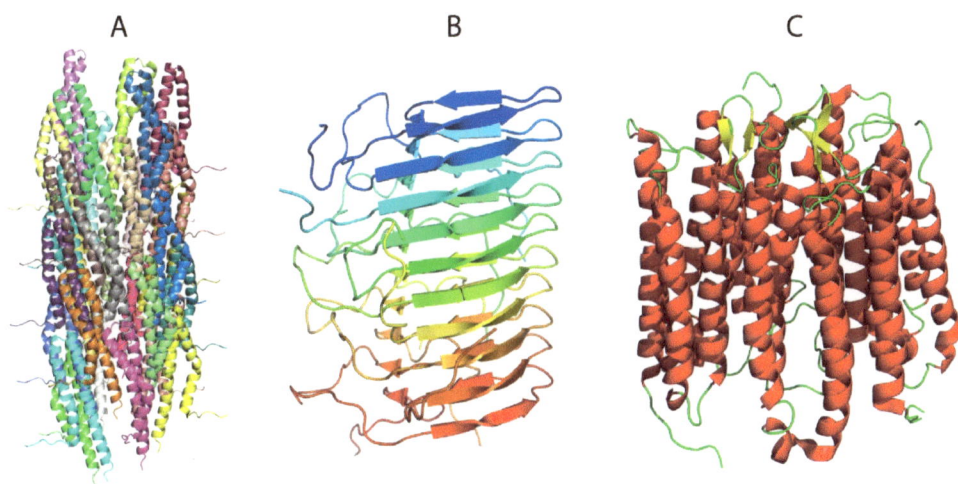

Fig. (10). Three examples of biomacromolecular structures solved with ssNMR data. **A)** The type-III secretion system needle from *Salmonella enterica* (ssNMR; PDB: 2LPZ) [40], **B)** The HET-s(218-289) prion protein of *Podospora anserina* (ssNMR; PDB: 2RNM) [99] and **C)** The *Anabaena* sensory rhodopsin membrane protein solubilised in DMPC-DMPA liposomes (ssNMR; ODB: 2M3G) [140]. Images were prepared in PyMOL [141] and are not to scale.

Fig. (**10B**) shows the HET-s functional amyloid prion domain from *Podospora anserina* [99]. The HET-s functional amyloid has become a model system for ssNMR due to its remarkable homogeneity and consequently well-resolved spectra. In this landmark work, the fibrillar structure of HET-s, with its β-solenoid fold was determined through ssNMR alone. From the analysis of spectra acquired using uniformly ^{13}C,^{15}N labelled sample, mixed ^{13}C and ^{15}N-labelled sample and spin-diluted 1,3-^{13}C-Glc and 2-^{13}C-Glc labelled samples, >100 ambiguous, specific inter-molecular, and specific intra-molecular restraints were identified. Furthermore, in addition to dihedral angle restraints defined from assigned chemical shifts, when secondary structure was in agreement with hydrogen-deuterium exchange data and proton-relayed spectra, specific hydrogen bond restraints were included.

Lastly, Fig. (**10C**) presents an example of a *de novo* integral membrane protein structure determined at atomic-resolution by ssNMR alone [140]. Prepared in DMPC-DMPA liposomes (to best mimic the native membrane environment), the homo-trimeric structure of the cyanobacterium *Anabaena* sensory rhodopsin (ASR) was determined primarily from the collection of >2000 total distance restraints, including ^{13}C-^{13}C and ^{1}H-^{1}H through-space correlations. Restraint assignment was facilitated by the acquisition of highly resolved spectra on both 1,3-^{13}C-Glc and 2-^{13}C-Glc labelled samples that were prepared in complex with the ^{13}C-labelled ASR cofactor retinal, for which intermolecular distance restraints were additionally identified. Interestingly, this study also involved the use of paramagnetic relaxation enhancement (PRE) restraints, modelled up to 15 Å, measured on a uniformly ^{13}C,^{15}N-labelled sample, with a single cysteine mutation in order to bind a paramagnetic nitroxide spin label, and such restraints helped to define the inter-subunit trimer interfaces.

CONCLUSION

A wide range of biophysical techniques is available to researchers for the study of biomolecular structures. The realm of protein crystallography has been explored for over 50 years and still offers the best achievable structural resolution for samples amenable to crystallisation. Likewise, solution NMR is the go-to method for the routine investigation of low molecular-weight samples (<50 kDa) in solution. However ssNMR has now emerged as a technique, which can readily probe the structural characteristics of biomolecular assemblies inaccessible to crystallography and solution NMR. These non-crystalline or insoluble species include gels, biofilms, membrane proteins, biological filaments, aggregates, capsids and protein assemblies under native (or near-native) conditions and at atomic-level resolution. Without always requiring the generation of full 3D models, ssNMR can provide informative, unique atomic insights into assembled biomolecular systems. Furthermore, the atomic data obtained from ssNMR is very complementary to other low- or high-resolution biophysical techniques, which can, for example, provide molecular envelopes or structures of soluble subunits.

The emergent technological advancements (hardware, spin diluting techniques, proton detection, structure calculation protocols) support that ssNMR has a real potential to fulfilling a key role in investigation on structure, interactions and dynamics of complex assembled biomolecular systems. However, it is to be expected that these complex targets for structural biology will become so very rarefied that they will begin to be intractable by any single approach. Consequently, integrative structural methods, which combine multiple structural data will start to become increasingly necessary and, amongst these techniques, ssNMR will remain key for its ability to probe these unique classes of macromolecular assemblies at atomic resolution.

CONSENT FOR PUBLICATION

Not applicable.

CONFLICT OF INTEREST

The authors declare no conflict of interest, financial or otherwise.

ACKNOWLEDGEMENTS

We acknowledge the financial support by the ANR (ANR-13-PDOC-0017-01 to B.H. and ANR-14-CE09-0020-01 to A.L.), the IdEx Bordeaux "Investments for the future" Chaire d'Installation and PEPS to B.H. (ANR-10-IDEX-03-02 to B.H.), the European Research Council (ERC-2015-StG GA no. 639020 to A.L.) and the CNRS (IR-RMN FR3050).

REFERENCES

[1]   Tycko, R. Molecular Structure of Aggregated Amyloid-beta: Insights from Solid-State Nuclear Magnetic Resonance. *Cold Spring Harb. Perspect. Med.,* **2016**, *6*.

[2]   van der Wel, P.C. Insights into protein misfolding and aggregation enabled by solid-state NMR spectroscopy. *Solid State Nucl. Magn. Reson.,* **2017**, *88*, 1-14.
[http://dx.doi.org/10.1016/j.ssnmr.2017.10.001]

[3]   Meier, B.H.; Riek, R.; Bockmann, A. Emerging Structural Understanding of Amyloid Fibrils by Solid-State NMR. *Trends Biochem. Sci.,* **2017**, *42*, 777-787.
[http://dx.doi.org/10.1016/j.tibs.2017.08.001]

[4]   Habenstein, B.; Loquet, A. Bacterial Filamentous Appendages Investigated by Solid-State NMR Spectroscopy. *Methods Mol. Biol.,* **2017**, *1615*, 415-448.
[http://dx.doi.org/10.1007/978-1-4939-7033-9_29]

[5]   Loquet, A.; Habenstein, B.; Lange, A. Structural investigations of molecular machines by solid-state NMR. *Acc. Chem. Res.,* **2013**, *46*, 2070-2079.
[http://dx.doi.org/10.1021/ar300320p]

[6]   Weingarth, M.; Baldus, M. Solid-state NMR-based approaches for supramolecular structure elucidation. *Acc. Chem. Res.,* **2013**, *46*, 2037-2046.
[http://dx.doi.org/10.1021/ar300316e]

[7]   Habenstein, B.; Loquet, A. Solid-state NMR: An emerging technique in structural biology of self-assemblies. *Biophys. Chem.,* **2016**, *210*, 14-26.
[http://dx.doi.org/10.1016/j.bpc.2015.07.003]

[8]   Linser, R. Solid-state NMR spectroscopic trends for supramolecular assemblies and protein aggregates. *Solid State Nucl. Magn. Reson.,* **2017**, *87*, 45-53.
[http://dx.doi.org/10.1016/j.ssnmr.2017.08.003]

[9]   Quinn, C.M.; Polenova, T. Structural biology of supramolecular assemblies by magic-angle spinning NMR spectroscopy. *Q. Rev. Biophys.,* **2017**, *50*, e1.
[http://dx.doi.org/10.1017/S0033583516000159]

[10]   Ullrich, S.J.; Glaubitz, C. Perspectives in enzymology of membrane proteins by solid-state NMR. *Acc.*

Chem. Res., **2013**, *46*, 2164-2171.
[http://dx.doi.org/10.1021/ar4000289]

[11]   Baker, L.A.; Baldus, M. Characterization of membrane protein function by solid-state NMR spectroscopy. *Curr. Opin. Struct. Biol.,* **2014**, *27*, 48-55.
[http://dx.doi.org/10.1016/j.sbi.2014.03.009]

[12]   Ladizhansky, V. Applications of solid-state NMR to membrane proteins *Biochim Biophys Acta,* **2017**, *1865*, 1577-86.
[http://dx.doi.org/10.1016/j.bbapap.2017.07.004]

[13]   Krushelnitsky, A.; Reichert, D.; Saalwachter, K. Solid-state NMR approaches to internal dynamics of proteins: from picoseconds to microseconds and seconds. *Acc. Chem. Res.,* **2013**, *46*, 2028-2036.
[http://dx.doi.org/10.1021/ar300292p]

[14]   Lewandowski, J.R. Advances in solid-state relaxation methodology for probing site-specific protein dynamics. *Acc. Chem. Res.,* **2013**, *46*, 2018-2027.
[http://dx.doi.org/10.1021/ar300334g]

[15]   Watt, E.D.; Rienstra, C.M. Recent advances in solid-state nuclear magnetic resonance techniques to quantify biomolecular dynamics. *Anal. Chem.,* **2014**, *86*, 58-64.
[http://dx.doi.org/10.1021/ac403956k]

[16]   Schanda, P.; Ernst, M. Studying Dynamics by Magic-Angle Spinning Solid-State NMR Spectroscopy: Principles and Applications to Biomolecules. *Prog. Nucl. Magn. Reson. Spectrosc.,* **2016**, *96*, 1-46.
[http://dx.doi.org/10.1016/j.pnmrs.2016.02.001]

[17]   Struppe, J.; Quinn, C.M.; Lu, M.; Wang, M.; Hou, G.; Lu, X. Expanding the horizons for structural analysis of fully protonated protein assemblies by NMR spectroscopy at MAS frequencies above 100 kHz. *Solid State Nucl. Magn. Reson.,* **2017**, *87*, 117-125.
[http://dx.doi.org/10.1016/j.ssnmr.2017.07.001]

[18]   Quinn, CM; Wang, M; Polenova, T NMR of Macromolecular Assemblies and Machines at 1 GHz and Beyond: New Transformative Opportunities for Molecular Structural Biology *Methods Mol Biol.,* **2018**, *1688*, 1-35.

[19]   Castellani, F.; van Rossum, B.; Diehl, A.; Schubert, M.; Rehbein, K.; Oschkinat, H. Structure of a protein determined by solid-state magic-angle-spinning NMR spectroscopy. *Nature,* **2002**, *420*, 98-102.
[http://dx.doi.org/10.1038/nature01070]

[20]   Loquet, A.; Tolchard, J.; Berbon, M.; Martinez, D.; Habenstein, B. Atomic Scale Structural Studies of Macromolecular Assemblies by Solid-state Nuclear Magnetic Resonance Spectroscopy. *J. Vis. Exp.,* **2017**.
[http://dx.doi.org/10.3791/55779]

[21]   Zhou, D.H.; Shah, G.; Cormos, M.; Mullen, C.; Sandoz, D.; Rienstra, C.M. Proton-detected solid-state NMR spectroscopy of fully protonated proteins at 40 kHz magic-angle spinning. *J. Am. Chem. Soc.,* **2007**, *129*, 11791-11801.
[http://dx.doi.org/10.1021/ja073462m]

[22]   Huber, M.; Hiller, S.; Schanda, P.; Ernst, M.; Bockmann, A.; Verel, R. A proton-detected 4D solid-state NMR experiment for protein structure determination. *Chem. Phys. Chem,* **2011**, *12*, 915-918.
[http://dx.doi.org/10.1002/cphc.201100062]

[23]   Knight, M.J.; Webber, A.L.; Pell, A.J.; Guerry, P.; Barbet-Massin, E.; Bertini, I. Fast resonance assignment and fold determination of human superoxide dismutase by high-resolution proton-detected solid-state MAS NMR spectroscopy. *Angew. Chem. Int. Ed. Engl.,* **2011**, *50*, 11697-11701.
[http://dx.doi.org/10.1002/anie.201106340]

[24]   Linser, R.; Bardiaux, B.; Higman, V.; Fink, U.; Reif, B. Structure calculation from unambiguous long-range amide and methyl 1H-1H distance restraints for a microcrystalline protein with MAS solid-state

NMR spectroscopy. *J. Am. Chem. Soc.,* **2011**, *133*, 5905-5912.
[http://dx.doi.org/10.1021/ja110222h]

[25]   Zhang, R.; Nishiyama, Y.; Ramamoorthy, A. Proton-detected 3D (1)H/(13)C/(1)H correlation experiment for structural analysis in rigid solids under ultrafast-MAS above 60 kHz. *J. Chem. Phys.,* **2015**, *143*, 164201.
[http://dx.doi.org/10.1063/1.4933373]

[26]   Penzel, S.; Smith, A.A.; Agarwal, V.; Hunkeler, A.; Org, M.L.; Samoson, A. Protein resonance assignment at MAS frequencies approaching 100 kHz: a quantitative comparison of J-coupling and dipolar-coupling-based transfer methods. *J. Biomol. NMR,* **2015**, *63*, 165-186.
[http://dx.doi.org/10.1007/s10858-015-9975-y]

[27]   Andreas, L.B.; Jaudzems, K.; Stanek, J.; Lalli, D.; Bertarello, A.; Le Marchand, T. Structure of fully protonated proteins by proton-detected magic-angle spinning NMR. *Proc. Natl. Acad. Sci. USA,* **2016**, *113*, 9187-9192.
[http://dx.doi.org/10.1073/pnas.1602248113]

[28]   Stanek, J.; Andreas, L.B.; Jaudzems, K.; Cala, D.; Lalli, D.; Bertarello, A. NMR Spectroscopic Assignment of Backbone and Side-Chain Protons in Fully Protonated Proteins: Microcrystals, Sedimented Assemblies, and Amyloid Fibrils. *Angew. Chem. Int. Ed. Engl.,* **2016**, *55*, 15504-15509.
[http://dx.doi.org/10.1002/anie.201607084]

[29]   Zhang, R.; Mroue, K.H.; Ramamoorthy, A. Proton-Based Ultrafast Magic Angle Spinning Solid-State NMR Spectroscopy. *Acc. Chem. Res.,* **2017**, *50*, 1105-1113.
[http://dx.doi.org/10.1021/acs.accounts.7b00082]

[30]   Zhang, R.; Duong, N.T.; Nishiyama, Y.; Ramamoorthy, A. 3D Double-Quantum/Double-Quantum Exchange Spectroscopy of Protons under 100 kHz Magic Angle Spinning. *J. Phys. Chem. B,* **2017**, *121*, 5944-5952.
[http://dx.doi.org/10.1021/acs.jpcb.7b03480]

[31]   Mroue, K.H.; Nishiyama, Y.; Kumar Pandey, M.; Gong, B.; McNerny, E.; Kohn, D.H. Proton-Detected Solid-State NMR Spectroscopy of Bone with Ultrafast Magic Angle Spinning. *Sci. Rep.,* **2015**, *5*, 11991.
[http://dx.doi.org/10.1038/srep11991]

[32]   Andreas, L.B.; Le Marchand, T.; Jaudzems, K.; Pintacuda, G. High-resolution proton-detected NMR of proteins at very fast MAS. *J. Magn. Reson.,* **2015**, *253*, 36-49.
[http://dx.doi.org/10.1016/j.jmr.2015.01.003]

[33]   Etzkorn, M.; Bockmann, A.; Lange, A.; Baldus, M. Probing molecular interfaces using 2D magic-angle-spinning NMR on protein mixtures with different uniform labeling. *J. Am. Chem. Soc.,* **2004**, *126*, 14746-14751.
[http://dx.doi.org/10.1021/ja0479181]

[34]   Loquet, A.; Giller, K.; Becker, S.; Lange, A. Supramolecular interactions probed by 13C-13C solid-state NMR spectroscopy. *J. Am. Chem. Soc.,* **2010**, *132*, 15164-15166.
[http://dx.doi.org/10.1021/ja107460j]

[35]   Shen, Y.; Delaglio, F.; Cornilescu, G.; Bax, A. TALOS+: a hybrid method for predicting protein backbone torsion angles from NMR chemical shifts. *J. Biomol. NMR,* **2009**, *44*, 213-223.
[http://dx.doi.org/10.1007/s10858-009-9333-z]

[36]   Berjanskii, M.V.; Neal, S.; Wishart, D.S. PREDITOR: a web server for predicting protein torsion angle restraints. *Nucleic Acids Res.,* **2006**, *34*, W63-9.
[http://dx.doi.org/10.1093/nar/gkl341]

[37]   Schwieters, C.D.; Kuszewski, J.J.; Tjandra, N.; Clore, G.M. The Xplor-NIH NMR molecular structure determination package. *J. Magn. Reson.,* **2003**, *160*, 65-73.
[http://dx.doi.org/10.1016/S1090-7807(02)00014-9]

[38]   Rieping, W.; Habeck, M.; Bardiaux, B.; Bernard, A.; Malliavin, T.E.; Nilges, M. ARIA2: automated NOE assignment and data integration in NMR structure calculation. *Bioinformatics,* **2007**, *23*, 381-382.
[http://dx.doi.org/10.1093/bioinformatics/btl589]

[39]   Herrmann, T.; Guntert, P.; Wuthrich, K. Protein NMR structure determination with automated NOE assignment using the new software CANDID and the torsion angle dynamics algorithm DYANA. *J. Mol. Biol.,* **2002**, *319*, 209-227.
[http://dx.doi.org/10.1016/S0022-2836(02)00241-3]

[40]   Loquet, A.; Sgourakis, N.G.; Gupta, R.; Giller, K.; Riedel, D.; Goosmann, C. Atomic model of the type III secretion system needle. *Nature,* **2012**, *486*, 276-279.

[41]   Habenstein, B.; Loquet, A.; Hwang, S.; Giller, K.; Vasa, S.K.; Becker, S. Hybrid Structure of the Type 1 Pilus of Uropathogenic Escherichia coli. *Angew. Chem. Int. Ed. Engl.,* **2015**, *54*, 11691-11695.
[http://dx.doi.org/10.1002/anie.201505065]

[42]   He, L.; Bardiaux, B.; Ahmed, M.; Spehr, J.; Konig, R.; Lunsdorf, H. Structure determination of helical filaments by solid-state NMR spectroscopy. *Proc. Natl. Acad. Sci. USA,* **2016**, *113*, E272-E281.
[http://dx.doi.org/10.1073/pnas.1513119113]

[43]   Demers, J.P.; Habenstein, B.; Loquet, A.; Kumar Vasa, S.; Giller, K.; Becker, S. High-resolution structure of the Shigella type-III secretion needle by solid-state NMR and cryo-electron microscopy. *Nat. Commun.,* **2014**, *5*, 4976.
[http://dx.doi.org/10.1038/ncomms5976]

[44]   Walti, M.A.; Ravotti, F.; Arai, H.; Glabe, C.G.; Wall, J.S.; Bockmann, A. Atomic-resolution structure of a disease-relevant Abeta(1-42) amyloid fibril. *Proc. Natl. Acad. Sci. USA,* **2016**, *113*, E4976-E4984.
[http://dx.doi.org/10.1073/pnas.1600749113]

[45]   Kotler, S.A.; Brender, J.R.; Vivekanandan, S.; Suzuki, Y.; Yamamoto, K.; Monette, M. High-resolution NMR characterization of low abundance oligomers of amyloid-beta without purification. *Sci. Rep.,* **2015**, *5*, 11811.
[http://dx.doi.org/10.1038/srep11811]

[46]   Agarwal, V.; Penzel, S.; Szekely, K.; Cadalbert, R.; Testori, E.; Oss, A. De novo 3D structure determination from sub-milligram protein samples by solid-state 100 kHz MAS NMR spectroscopy. *Angew. Chem. Int. Ed. Engl.,* **2014**, *53*, 12253-12256.
[http://dx.doi.org/10.1002/anie.201405730]

[47]   Nishiyama, Y.; Malon, M.; Ishii, Y.; Ramamoorthy, A., III (1)(5)N/(1)(5)N/(1)H chemical shift correlation experiment utilizing an RFDR-based (1)H/(1)H mixing period at 100 kHz MAS. *J. Magn. Reson.,* **2014**, *244*, 1-5.
[http://dx.doi.org/10.1016/j.jmr.2014.04.008]

[48]   Yamamoto, K; Caporini, MA Cellular solid-state NMR investigation of a membrane protein using dynamic nuclear polarization *Biochim Biophys Acta,* **2015**, *1848*, 342-9.

[49]   Yamamoto, K.; Caporini, M.A.; Im, S.C.; Waskell, L.; Ramamoorthy, A. Transmembrane Interactions of Full-length Mammalian Bitopic Cytochrome-P450-Cytochrome-b5 Complex in Lipid Bilayers Revealed by Sensitivity-Enhanced Dynamic Nuclear Polarization Solid-state NMR Spectroscopy. *Sci. Rep.,* **2017**, *7*, 4116.
[http://dx.doi.org/10.1038/s41598-017-04219-1]

[50]   Fricke, P.; Demers, J.P.; Becker, S.; Lange, A. Studies on the MxiH protein in T3SS needles using DNP-enhanced ssNMR spectroscopy. *ChemPhysChem,* **2014**, *15*, 57-60.
[http://dx.doi.org/10.1002/cphc.201300994]

[51]   Kaplan, M.; Cukkemane, A.; van Zundert, G.C.; Narasimhan, S.; Daniels, M.; Mance, D. Probing a cell-embedded megadalton protein complex by DNP-supported solid-state NMR. *Nat. Methods,* **2015**, *12*, 649-652.

[http://dx.doi.org/10.1038/nmeth.3406]

[52]   Bertini, I.; Luchinat, C.; Parigi, G.; Ravera, E.; Reif, B.; Turano, P. Solid-state NMR of proteins sedimented by ultracentrifugation. *Proc. Natl. Acad. Sci. USA,* **2011**, *108*, 10396-10399.
        [http://dx.doi.org/10.1073/pnas.1103854108]

[53]   Durr, UH; Waskell, L; Ramamoorthy, A The cytochromes P450 and b5 and their reductases--promising targets for structural studies by advanced solid-state NMR spectroscopy *Biochim Biophys Acta,* **2007**, *1768*, 3235-59.

[54]   Opella, S.J.; Marassi, F.M. Applications of NMR to membrane proteins. *Arch. Biochem. Biophys.,* **2017**, *628*, 92-101.
        [http://dx.doi.org/10.1016/j.abb.2017.05.011]

[55]   Wang, S.; Ladizhansky, V. Recent advances in magic angle spinning solid state NMR of membrane proteins. *Prog. Nucl. Magn. Reson. Spectrosc.,* **2014**, *82*, 1-26.
        [http://dx.doi.org/10.1016/j.pnmrs.2014.07.001]

[56]   Durr, U.H.; Gildenberg, M.; Ramamoorthy, A. The magic of bicelles lights up membrane protein structure. *Chem. Rev.,* **2012**, *112*, 6054-6074.
        [http://dx.doi.org/10.1021/cr300061w]

[57]   Durr, U.H.; Soong, R.; Ramamoorthy, A. When detergent meets bilayer: birth and coming of age of lipid bicelles. *Prog. Nucl. Magn. Reson. Spectrosc.,* **2013**, *69*, 1-22.
        [http://dx.doi.org/10.1016/j.pnmrs.2013.01.001]

[58]   Hallock, K.J.; Lee, D.K.; Ramamoorthy, A. MSI-78, an analogue of the magainin antimicrobial peptides, disrupts lipid bilayer structure *via* positive curvature strain. *Biophys. J.,* **2003**, *84*, 3052-3060.
        [http://dx.doi.org/10.1016/S0006-3495(03)70031-9]

[59]   Park, S.H.; Das, B.B.; Casagrande, F.; Tian, Y.; Nothnagel, H.J.; Chu, M. Structure of the chemokine receptor CXCR1 in phospholipid bilayers. *Nature,* **2012**, *491*, 779-783.

[60]   Murray, D.T.; Griffin, J.; Cross, T.A. Detergent optimized membrane protein reconstitution in liposomes for solid state NMR. *Biochemistry,* **2014**, *53*, 2454-2463.
        [http://dx.doi.org/10.1021/bi500144h]

[61]   Park, S.H.; Berkamp, S.; Cook, G.A.; Chan, M.K.; Viadiu, H.; Opella, S.J. Nanodiscs versus macrodiscs for NMR of membrane proteins. *Biochemistry,* **2011**, *50*, 8983-8985.
        [http://dx.doi.org/10.1021/bi201289c]

[62]   Ding, Y.; Fujimoto, L.M.; Yao, Y.; Marassi, F.M. Solid-state NMR of the Yersinia pestis outer membrane protein Ail in lipid bilayer nanodiscs sedimented by ultracentrifugation. *J. Biomol. NMR,* **2015**, *61*, 275-286.
        [http://dx.doi.org/10.1007/s10858-014-9893-4]

[63]   Ravula, T.; Hardin, N.Z.; Ramadugu, S.; Cox, S.J.; Ramamoorthy, A. pH resistant monodispersed polymer-lipid nanodiscs. *Angew. Chem. Int. Ed. Engl.,* **2017**.

[64]   Baker, LA; Sinnige, T; Schellenberger, P; de Keyzer, J; Siebert, CA; Driessen, AJM; Baldus, M; Grünewald, K. Combined (1)H-detected solid-state NMR spectroscopy and electron cryotomography to study membrane proteins across resolutions in native environments. *Structure,* **2018**, *26*, 161-70 e3.

[65]   Brown, L.S.; Ladizhansky, V. Membrane proteins in their native habitat as seen by solid-state NMR spectroscopy. *Protein Sci.,* **2015**, *24*, 1333-1346.
        [http://dx.doi.org/10.1002/pro.2700]

[66]   Stevens, R.C. Design of high-throughput methods of protein production for structural biology. *Structure,* **2000**, *8*, R177-R185.
        [http://dx.doi.org/10.1016/S0969-2126(00)00193-3]

[67]   Strausberg, R.L.; Strausberg, S.L. Overview of protein expression in Saccharomyces cerevisiae In: *Curr. Protoc. Protein. Sci*; , **2001**; Chapter 5, p. Unit5.6.

[68] Nettleship, J.E.; Assenberg, R.; Diprose, J.M.; Rahman-Huq, N.; Owens, R.J. Recent advances in the production of proteins in insect and mammalian cells for structural biology. *J. Struct. Biol.*, **2010**, *172*, 55-65.
[http://dx.doi.org/10.1016/j.jsb.2010.02.006]

[69] Portolano, N.; Watson, P.J.; Fairall, L.; Millard, C.J.; Milano, C.P.; Song, Y. Recombinant protein expression for structural biology in HEK 293F suspension cells: a novel and accessible approach. *J. Vis. Exp.*, **2014**, e51897.

[70] Vinarov, D.A.; Newman, C.L.; Tyler, E.M.; Markley, J.L.; Shahan, M.N. Wheat germ cell-free expression system for protein production. In: *Curr. Protoc. Protein. Sci*; , **2006**; Chapter 5, p. Unit 5.18.
[http://dx.doi.org/10.1002/0471140864.ps0518s44]

[71] Kigawa, T.; Yabuki, T.; Yoshida, Y.; Tsutsui, M.; Ito, Y.; Shibata, T. Cell-free production and stable-isotope labeling of milligram quantities of proteins. *FEBS Lett.*, **1999**, *442*, 15-19.
[http://dx.doi.org/10.1016/S0014-5793(98)01620-2]

[72] Rosano, G.L.; Ceccarelli, E.A. Recombinant protein expression in Escherichia coli: advances and challenges. *Front. Microbiol.*, **2014**, *5*, 172.
[http://dx.doi.org/10.3389/fmicb.2014.00172]

[73] Studier, F.W. Protein production by auto-induction in high density shaking cultures. *Protein Expr. Purif.*, **2005**, *41*, 207-234.
[http://dx.doi.org/10.1016/j.pep.2005.01.016]

[74] Cai, M.; Huang, Y.; Yang, R.; Craigie, R.; Clore, G.M. A simple and robust protocol for high-yield expression of perdeuterated proteins in Escherichia coli grown in shaker flasks. *J. Biomol. NMR*, **2016**, *66*, 85-91.
[http://dx.doi.org/10.1007/s10858-016-0052-y]

[75] Reif, B.; Jaroniec, C.P.; Rienstra, C.M.; Hohwy, M.; Griffin, R.G. 1H-1H MAS correlation spectroscopy and distance measurements in a deuterated peptide. *J. Magn. Reson.*, **2001**, *151*, 320-327.
[http://dx.doi.org/10.1006/jmre.2001.2354]

[76] Paulson, E.K.; Morcombe, C.R.; Gaponenko, V.; Dancheck, B.; Byrd, R.A.; Zilm, K.W. Sensitive high resolution inverse detection NMR spectroscopy of proteins in the solid state. *J. Am. Chem. Soc.*, **2003**, *125*, 15831-15836.
[http://dx.doi.org/10.1021/ja037315+]

[77] Venters, R.A.; Huang, C.C.; Farmer, B.T., II; Trolard, R.; Spicer, L.D.; Fierke, C.A. High-level 2H/13C/15N labeling of proteins for NMR studies. *J. Biomol. NMR*, **1995**, *5*, 339-344.
[http://dx.doi.org/10.1007/BF00182275]

[78] Linser, R.; Fink, U.; Reif, B. Proton-detected scalar coupling based assignment strategies in MAS solid-state NMR spectroscopy applied to perdeuterated proteins. *J. Magn. Reson.*, **2008**, *193*, 89-93.
[http://dx.doi.org/10.1016/j.jmr.2008.04.021]

[79] Akbey, U.; Lange, S.; Trent Franks, W.; Linser, R.; Rehbein, K.; Diehl, A. Optimum levels of exchangeable protons in perdeuterated proteins for proton detection in MAS solid-state NMR spectroscopy. *J. Biomol. NMR*, **2010**, *46*, 67-73.
[http://dx.doi.org/10.1007/s10858-009-9369-0]

[80] Linser, R.; Fink, U.; Reif, B. Narrow carbonyl resonances in proton-diluted proteins facilitate NMR assignments in the solid-state. *J. Biomol. NMR*, **2010**, *47*, 1-6.
[http://dx.doi.org/10.1007/s10858-010-9404-1]

[81] Linser, R.; Fink, U.; Reif, B. Assignment of dynamic regions in biological solids enabled by spin-state selective NMR experiments. *J. Am. Chem. Soc.*, **2010**, *132*, 8891-8893.
[http://dx.doi.org/10.1021/ja102612m]

[82]   Linser, R. Backbone assignment of perdeuterated proteins using long-range H/C-dipolar transfers. *J. Biomol. NMR,* **2012**, *52*, 151-158.
[http://dx.doi.org/10.1007/s10858-011-9593-2]

[83]   Fasshuber, H.K.; Demers, J.P.; Chevelkov, V.; Giller, K.; Becker, S.; Lange, A. Specific 13C labeling of leucine, valine and isoleucine methyl groups for unambiguous detection of long-range restraints in protein solid-state NMR studies. *J. Magn. Reson.,* **2015**, *252*, 10-19.
[http://dx.doi.org/10.1016/j.jmr.2014.12.013]

[84]   Lin, M.T.; Fukazawa, R.; Miyajima-Nakano, Y.; Matsushita, S.; Choi, S.K.; Iwasaki, T. Escherichia coli auxotroph host strains for amino acid-selective isotope labeling of recombinant proteins. *Methods Enzymol.,* **2015**, *565*, 45-66.
[http://dx.doi.org/10.1016/bs.mie.2015.05.012]

[85]   Noirot, C.; Habenstein, B.; Bousset, L.; Melki, R.; Meier, B.H.; Endo, Y. Wheat-germ cell-free production of prion proteins for solid-state NMR structural studies. *N. Biotechnol.,* **2011**, *28*, 232-238.
[http://dx.doi.org/10.1016/j.nbt.2010.06.016]

[86]   Abdine, A.; Park, K.H.; Warschawski, D.E. Cell-free membrane protein expression for solid-state NMR. *Methods Mol. Biol.,* **2012**, *831*, 85-109.
[http://dx.doi.org/10.1007/978-1-61779-480-3_6]

[87]   Fogeron, M.L.; Jirasko, V.; Penzel, S.; Paul, D.; Montserret, R.; Danis, C. Cell-free expression, purification, and membrane reconstitution for NMR studies of the nonstructural protein 4B from hepatitis C virus. *J. Biomol. NMR,* **2016**, *65*, 87-98.
[http://dx.doi.org/10.1007/s10858-016-0040-2]

[88]   Singh, A.; Upadhyay, V.; Upadhyay, A.K.; Singh, S.M.; Panda, A.K. Protein recovery from inclusion bodies of Escherichia coli using mild solubilization process. *Microb. Cell Fact.,* **2015**, *14*, 41.
[http://dx.doi.org/10.1186/s12934-015-0222-8]

[89]   McManus, J.J.; Charbonneau, P.; Zaccarelli, E.; Asherie, N. The physics of protein self-assembly. *Curr. Opin. Colloid Interface Sci.,* **2016**, *22*, 73-79.
[http://dx.doi.org/10.1016/j.cocis.2016.02.011]

[90]   Serio, T.R.; Cashikar, A.G.; Kowal, A.S.; Sawicki, G.J.; Moslehı, J.J.; Serpell, L. Nucleated conformational conversion and the replication of conformational information by a prion determinant. *Science,* **2000**, *289*, 1317-1321.
[http://dx.doi.org/10.1126/science.289.5483.1317]

[91]   Svane, A.S.; Jahn, K.; Deva, T.; Malmendal, A.; Otzen, D.E.; Dittmer, J. Early stages of amyloid fibril formation studied by liquid-state NMR: the peptide hormone glucagon. *Biophys. J.,* **2008**, *95*, 366-377.
[http://dx.doi.org/10.1529/biophysj.107.122895]

[92]   Chan, F.T.; Kaminski Schierle, G.S.; Kumita, J.R.; Bertoncini, C.W.; Dobson, C.M.; Kaminski, C.F. Protein amyloids develop an intrinsic fluorescence signature during aggregation. *Analyst (Lond.),* **2013**, *138*, 2156-2162.
[http://dx.doi.org/10.1039/c3an36798c]

[93]   Pardee, J.D.; Spudich, J.A. Purification of muscle actin. *Struct Contract Proteins Part B Contract Appar Cytoskelet.,* **1982**, *85*, 164-181.
[http://dx.doi.org/10.1016/0076-6879(82)85020-9]

[94]   Bockmann, A.; Gardiennet, C.; Verel, R.; Hunkeler, A.; Loquet, A.; Pintacuda, G. Characterization of different water pools in solid-state NMR protein samples. *J. Biomol. NMR,* **2009**, *45*, 319-327.
[http://dx.doi.org/10.1007/s10858-009-9374-3]

[95]   Bertini, I.; Engelke, F.; Gonnelli, L.; Knott, B.; Luchinat, C.; Osen, D. On the use of ultracentrifugal devices for sedimented solute NMR. *J. Biomol. NMR,* **2012**, *54*, 123-127.
[http://dx.doi.org/10.1007/s10858-012-9657-y]

[96]   Hisao, G.S.; Harland, M.A.; Brown, R.A.; Berthold, D.A.; Wilson, T.E.; Rienstra, C.M. An efficient

method and device for transfer of semisolid materials into solid-state NMR spectroscopy rotors. *J. Magn. Reson.,* **2016**, *265*, 172-176.
[http://dx.doi.org/10.1016/j.jmr.2016.01.009]

[97]   Cavanagh, J; Fairbrother, WJ; Palmer, AG; Skelton, NJ Protein NMR spectroscopy: Principles and Practice **1996**.

[98]   Daskalov, A.; Habenstein, B.; Sabate, R.; Berbon, M.; Martinez, D.; Chaignepain, S. Identification of a novel cell death-inducing domain reveals that fungal amyloid-controlled programmed cell death is related to necroptosis. *Proc. Natl. Acad. Sci. USA,* **2016**, *113*, 2720-2725.
[http://dx.doi.org/10.1073/pnas.1522361113]

[99]   Wasmer, C.; Lange, A.; Van Melckebeke, H.; Siemer, A.B.; Riek, R.; Meier, B.H. Amyloid fibrils of the HET-s(218-289) prion form a beta solenoid with a triangular hydrophobic core. *Science,* **2008**, *319*, 1523-1526.
[http://dx.doi.org/10.1126/science.1151839]

[100] Szeverenyi, N.M.; Sullivan, M.J.; Maciel, G.E. Observation of spin exchange by two-dimensional fourier transform 13C cross polarization-magic-angle spinning. *J. Magn. Reson.,* **1982**, *47*, 462-475.

[101] Wang, Y.; Jardetzky, O. Probability-based protein secondary structure identification using combined NMR chemical-shift data. *Protein Sci.,* **2002**, *11*, 852-861.
[http://dx.doi.org/10.1110/ps.3180102]

[102] Hafsa, N.E.; Arndt, D.; Wishart, D.S. CSI 3.0: a web server for identifying secondary and super-secondary structure in proteins using NMR chemical shifts. *Nucleic Acids Res.,* **2015**, *43*, W370-7.
[http://dx.doi.org/10.1093/nar/gkv494]

[103] Ulrich, E.L.; Akutsu, H.; Doreleijers, J.F.; Harano, Y.; Ioannidis, Y.E.; Lin, J. BioMagResBank. *Nucleic Acids Res.,* **2008**, *36*, D402-D408.
[http://dx.doi.org/10.1093/nar/gkm957]

[104] Vranken, W.F.; Boucher, W.; Stevens, T.J.; Fogh, R.H.; Pajon, A.; Llinas, M. The CCPN data model for NMR spectroscopy: development of a software pipeline. *Proteins,* **2005**, *59*, 687-696.
[http://dx.doi.org/10.1002/prot.20449]

[105] Han, B.; Liu, Y.; Ginzinger, S.W.; Wishart, D.S. SHIFTX2: significantly improved protein chemical shift prediction. *J. Biomol. NMR,* **2011**, *50*, 43-57.
[http://dx.doi.org/10.1007/s10858-011-9478-4]

[106] Buchan, D.W.; Minneci, F.; Nugent, T.C.; Bryson, K.; Jones, D.T. Scalable web services for the PSIPRED Protein Analysis Workbench. *Nucleic Acids Res.,* **2013**, *41*, W349-57.
[http://dx.doi.org/10.1093/nar/gkt381]

[107] Verel, R.; Baldus, M.; Ernst, M.; Meier, B. A homonuclear spin-pair filter for solid-state NMR based on adiabatic-passage techniques. *Chem. Phys. Lett.,* **1998**, *287*, 421-428.
[http://dx.doi.org/10.1016/S0009-2614(98)00172-9]

[108] Verel, R.; Ernst, M.; Meier, B.H. Adiabatic dipolar recoupling in solid-state NMR: the DREAM scheme. *J. Magn. Reson.,* **2001**, *150*, 81-99.
[http://dx.doi.org/10.1006/jmre.2001.2310]

[109] Takegoshi, K.; Nakamura, S.; Terao, T. 13C–1H dipolar-assisted rotational resonance in magic-angle spinning NMR. *Chem. Phys. Lett.,* **2001**, *344*, 631-637.
[http://dx.doi.org/10.1016/S0009-2614(01)00791-6]

[110] Pauli, J.; Baldus, M.; van Rossum, B.; de Groot, H.; Oschkinat, H. Backbone and side-chain 13C and 15N signal assignments of the alpha-spectrin SH3 domain by magic angle spinning solid-state NMR at 17.6 Tesla. *ChemBioChem,* **2001**, *2*, 272-281.
[http://dx.doi.org/10.1002/1439-7633(20010401)2:4<272::AID-CBIC272>3.0.CO;2-2]

[111] Li, Y.; Berthold, D.A.; Frericks, H.L.; Gennis, R.B.; Rienstra, C.M. Partial (13)C and (15)N chemical-shift assignments of the disulfide-bond-forming enzyme DsbB by 3D magic-angle spinning NMR

spectroscopy. *ChemBioChem,* **2007**, *8*, 434-442.
[http://dx.doi.org/10.1002/cbic.200600484]

[112] Ward, M.E.; Shi, L.; Lake, E.; Krishnamurthy, S.; Hutchins, H.; Brown, L.S. Proton-detected solid-state NMR reveals intramembrane polar networks in a seven-helical transmembrane protein proteorhodopsin. *J. Am. Chem. Soc.,* **2011**, *133*, 17434-17443.
[http://dx.doi.org/10.1021/ja207137h]

[113] Zhou, D.H.; Nieuwkoop, A.J.; Berthold, D.A.; Comellas, G.; Sperling, L.J.; Tang, M. Solid-state NMR analysis of membrane proteins and protein aggregates by proton detected spectroscopy. *J. Biomol. NMR,* **2012**, *54*, 291-305.
[http://dx.doi.org/10.1007/s10858-012-9672-z]

[114] Barbet-Massin, E.; Pell, A.J.; Retel, J.S.; Andreas, L.B.; Jaudzems, K.; Franks, W.T. Rapid proton-detected NMR assignment for proteins with fast magic angle spinning. *J. Am. Chem. Soc.,* **2014**, *136*, 12489-12497.
[http://dx.doi.org/10.1021/ja507382j]

[115] Chevelkov, V.; Habenstein, B.; Loquet, A.; Giller, K.; Becker, S.; Lange, A. Proton-detected MAS NMR experiments based on dipolar transfers for backbone assignment of highly deuterated proteins. *J. Magn. Reson.,* **2014**, *242*, 180-188.
[http://dx.doi.org/10.1016/j.jmr.2014.02.020]

[116] Xiang, S.; Biernat, J.; Mandelkow, E.; Becker, S.; Linser, R. Backbone assignment for minimal protein amounts of low structural homogeneity in the absence of deuteration. *Chem. Commun. (Camb.),* **2016**, *52*, 4002-4005.
[http://dx.doi.org/10.1039/C5CC09160H]

[117] Vasa, S.K.; Rovo, P.; Giller, K.; Becker, S.; Linser, R. Access to aliphatic protons as reporters in non-deuterated proteins by solid-state NMR. *Phys. Chem. Chem. Phys.,* **2016**, *18*, 8359-8363.
[http://dx.doi.org/10.1039/C5CP06601H]

[118] Hardy, E.H.; Verel, R.; Meier, B.H. Fast MAS total through-bond correlation spectroscopy. *J. Magn. Reson.,* **2001**, *148*, 459-464.
[http://dx.doi.org/10.1006/jmre.2000.2258]

[119] Andronesi, O.C.; Becker, S.; Seidel, K.; Heise, H.; Young, H.S.; Baldus, M. Determination of membrane protein structure and dynamics by magic-angle-spinning solid-state NMR spectroscopy. *J. Am. Chem. Soc.,* **2005**, *127*, 12965-12974.
[http://dx.doi.org/10.1021/ja0530164]

[120] Andronesi, O.C.; von Bergen, M.; Biernat, J.; Seidel, K.; Griesinger, C.; Mandelkow, E. Characterization of Alzheimer's-like paired helical filaments from the core domain of tau protein using solid-state NMR spectroscopy. *J. Am. Chem. Soc.,* **2008**, *130*, 5922-5928.
[http://dx.doi.org/10.1021/ja7100517]

[121] Cheung, M.S.; Maguire, M.L.; Stevens, T.J.; Broadhurst, R.W. DANGLE: A Bayesian inferential method for predicting protein backbone dihedral angles and secondary structure. *J. Magn. Reson.,* **2010**, *202*, 223-233.
[http://dx.doi.org/10.1016/j.jmr.2009.11.008]

[122] Cavalli, A.; Salvatella, X.; Dobson, C.M.; Vendruscolo, M. Protein structure determination from NMR chemical shifts. *Proc. Natl. Acad. Sci. USA,* **2007**, *104*, 9615-9620.
[http://dx.doi.org/10.1073/pnas.0610313104]

[123] Shen, Y.; Lange, O.; Delaglio, F.; Rossi, P.; Aramini, J.M.; Liu, G. Consistent blind protein structure generation from NMR chemical shift data. *Proc. Natl. Acad. Sci. USA,* **2008**, *105*, 4685-4690.
[http://dx.doi.org/10.1073/pnas.0800256105]

[124] Robustelli, P.; Cavalli, A.; Vendruscolo, M. Determination of protein structures in the solid state from NMR chemical shifts. *Structure,* **2008**, *16*, 1764-1769.
[http://dx.doi.org/10.1016/j.str.2008.10.016]

[125] Skora, L.; Zweckstetter, M. Determination of amyloid core structure using chemical shifts. *Protein Sci.*, **2012**, *21*, 1948-1953.
[http://dx.doi.org/10.1002/pro.2170]

[126] Lange, A.; Luca, S.; Baldus, M. Structural constraints from proton-mediated rare-spin correlation spectroscopy in rotating solids. *J. Am. Chem. Soc.*, **2002**, *124*, 9704-9705.
[http://dx.doi.org/10.1021/ja026691b]

[127] Habenstein, B.; Loquet, A.; Giller, K.; Becker, S.; Lange, A. Structural characterization of supramolecular assemblies by (1)(3)C spin dilution and 3D solid-state NMR. *J. Biomol. NMR*, **2013**, *55*, 1-9.
[http://dx.doi.org/10.1007/s10858-012-9691-9]

[128] Nederveen, A.J.; Doreleijers, J.F.; Vranken, W.; Miller, Z.; Spronk, C.A.; Nabuurs, S.B. RECOORD: a recalculated coordinate database of 500+ proteins from the PDB using restraints from the BioMagResBank. *Proteins*, **2005**, *59*, 662-672.
[http://dx.doi.org/10.1002/prot.20408]

[129] Brunger, A.T.; Adams, P.D.; Clore, G.M.; DeLano, W.L.; Gros, P.; Grosse-Kunstleve, R.W. Crystallography & NMR system: A new software suite for macromolecular structure determination. *Acta Crystallogr. D Biol. Crystallogr.*, **1998**, *54*, 905-921.
[http://dx.doi.org/10.1107/S0907444998003254]

[130] Guntert, P. Automated NMR structure calculation with CYANA. *Methods Mol. Biol.*, **2004**, *278*, 353-378.

[131] Jehle, S.; Vollmar, B.S.; Bardiaux, B.; Dove, K.K.; Rajagopal, P.; Gonen, T. N-terminal domain of alphaB-crystallin provides a conformational switch for multimerization and structural heterogeneity. *Proc. Natl. Acad. Sci. USA*, **2011**, *108*, 6409-6414.
[http://dx.doi.org/10.1073/pnas.1014656108]

[132] Verardi, R.; Shi, L.; Traaseth, N.J.; Walsh, N.; Veglia, G. Structural topology of phospholamban pentamer in lipid bilayers by a hybrid solution and solid-state NMR method. *Proc. Natl. Acad. Sci. USA*, **2011**, *108*, 9101-9106.
[http://dx.doi.org/10.1073/pnas.1016535108]

[133] Sborgi, L.; Ravotti, F.; Dandey, V.P.; Dick, M.S.; Mazur, A.; Reckel, S. Structure and assembly of the mouse ASC inflammasome by combined NMR spectroscopy and cryo-electron microscopy. *Proc. Natl. Acad. Sci. USA*, **2015**, *112*, 13237-13242.
[http://dx.doi.org/10.1073/pnas.1507579112]

[134] Manolikas, T.; Herrmann, T.; Meier, B.H. Protein structure determination from 13C spin-diffusion solid-state NMR spectroscopy. *J. Am. Chem. Soc.*, **2008**, *130*, 3959-3966.
[http://dx.doi.org/10.1021/ja078039s]

[135] Bhattacharya, A.; Tejero, R.; Montelione, G.T. Evaluating protein structures determined by structural genomics consortia. *Proteins*, **2007**, *66*, 778-795.
[http://dx.doi.org/10.1002/prot.21165]

[136] Vriend, G. WHAT IF: a molecular modeling and drug design program. J Mol Graph. 1990; 8: 52-6, 29.

[137] Davis, I.W.; Leaver-Fay, A.; Chen, V.B.; Block, J.N.; Kapral, G.J.; Wang, X. MolProbity: all-atom contacts and structure validation for proteins and nucleic acids. *Nucleic Acids Res.*, **2007**, *35*, W375-83.
[http://dx.doi.org/10.1093/nar/gkm216]

[138] Das, R.; Andre, I.; Shen, Y.; Wu, Y.; Lemak, A.; Bansal, S. Simultaneous prediction of protein folding and docking at high resolution. *Proc. Natl. Acad. Sci. USA*, **2009**, *106*, 18978-18983.
[http://dx.doi.org/10.1073/pnas.0904407106]

[139] Fujii, T.; Cheung, M.; Blanco, A.; Kato, T.; Blocker, A.J.; Namba, K. Structure of a type III secretion

needle at 7-A resolution provides insights into its assembly and signaling mechanisms. *Proc. Natl. Acad. Sci. USA,* **2012**, *109*, 4461-4466.
[http://dx.doi.org/10.1073/pnas.1116126109]

[140] Wang, S.; Munro, R.A.; Shi, L.; Kawamura, I.; Okitsu, T.; Wada, A. Solid-state NMR spectroscopy structure determination of a lipid-embedded heptahelical membrane protein. *Nat. Methods,* **2013**, *10*, 1007-1012.
[http://dx.doi.org/10.1038/nmeth.2635]

[141] DeLano, WL. The PyMOL Molecular Graphics System.

CHAPTER 2

Assessing Liver Disease and the Gut Microbiome by NMR Metabolic Profiling of Body Fluids

I. Jane Cox[1,2,*], Mark J.W. McPhail[3] and Roger Williams[1,2]

[1] *Institute of Hepatology, London, Foundation for Liver Research, 111 Coldharbour Lane, London SE5 9NT, United Kingdom*

[2] *Faculty of Life Sciences & Medicine, King's College London, London SE5 9RS, United Kingdom*

[3] *Institute of Liver Sciences, Department of Inflammation Biology, School of Immunology and Microbial Sciences, Faculty of Life Sciences and Medicine, King's College London, London SE5 9RS, United Kingdom*

Abstract: The initial stages of liver damage can be difficult to detect using standard clinical and imaging diagnostic tests. Prior to the development of advanced fibrosis or liver failure, the diseased liver may abnormally metabolise nutrients and drugs. Such changes can be measured by differences in low molecular weight metabolites in body fluids using a range of state-or-the-art analytical chemistry methods, including proton nuclear magnetic resonance spectroscopy and mass spectrometry. Gut microbial co-metabolites, for example hippurate and trimethylamine-*N*-oxide, may also be detected in urine and blood. In this chapter we illustrate results of urinary, plasma and serum metabolic profiling, using proton nuclear magnetic resonance spectroscopy, for characterising specific aspects of liver disease and monitoring treatment of liver cirrhosis.

Keywords: Cirrhosis, Dimethylamine, Gut microbiome, Hepatocellular carcinoma, Hippurate, Liver disease, Nuclear magnetic resonance spectroscopy, Phenylacetylglutamine, Trimethylamine, Trimethylamine-*N*-oxide.

INTRODUCTION

The Liver and Gut Microbiota

The liver, as the largest organ in the body and the principal homeostatic organ, is considered to have up to 500 functions and processes, often in combination with other systems and organs. The major functions of the liver include: the

---

* **Corresponding author I. Jane Cox:** Institute of Hepatology, London, Foundation for Liver Research, 111 Coldharbour Lane, London SE5 9NT, UK, Tel: +44 (0)20 7255 9830; Fax: +44(0) +44 (0)20 7380 0405; E-mail: j.cox@researchinliver.org.uk

Atta-ur-Rahman & M. Iqbal Choudhary (Eds.)

metabolism of carbohydrates, fats and circulating proteins; the detoxification of the body's waste products; the most important site for drug and alcohol metabolism; a site of complex immunological activity; as well as generating a range of hormones, enzymes, and blood clotting factors. Bile is produced in the liver, for the digestion of fats, and the transport and excretion of bilirubin and certain drugs.

The liver is continually exposed, via the portal vein to venous outflow of the intestine, to circulating gut-derived factors, including bacterial metabolites and bacterial components [1, 2]. A range of liver disorders, including alcoholic liver disease, non-alcoholic fatty liver disease and primary sclerosing cholangitis, have been associated with pathological alteration of the gut microbiome [2, 3]. This "dysbiosis" may influence the degree of hepatic steatosis, inflammation and fibrosis through multiple interactions with the host's immune system and other cell types. A better understanding of the influence of gut microbiota in liver diseases is required as a basis for the development of novel therapies.

The Clinical Need for the Earlier Detection of Liver Disease

A rapid method for detecting the initial stages of liver damage in high risk subjects or the early signs of liver cancer on a background of cirrhosis is an important need, as these can be difficult to detect using routine biochemical and imaging diagnostic procedures [4]. Liver function tests include measurement of liver enzymes, blood clotting time and levels of albumin and bilirubin. The liver typically responds to injury by releasing enzymes from hepatocytes and/or the biliary epithelium, so elevated levels of aspartate aminotransferase, alanine aminotransferase and alkaline phosphatase are characteristically found. However, elevations of these enzymes do not necessarily correlate with the severity of liver injury, particularly as some enzymes are also expressed in other organs.

Imaging techniques play an important role in evaluating liver disease, but each of ultrasound, ultrasound elastography, X-ray computer tomography, and all the various magnetic resonance imaging methodologies, have limitations.

Histology results from liver biopsy remain the gold standard for the diagnosis and staging of liver disease, even though only $1/50,000^{th}$ of the entire liver is sampled by a single biopsy. Furthermore, whilst complications are rare from the procedure they can be serious or fatal, and repeated biopsies are not deemed acceptable to many patients.

The need for a simple and robust diagnostic test therefore remains. As the diseased liver may cause early changes in how nutrients or drugs are metabolised such differences may be detected by changes in low molecular weight metabolites in either blood or urine [5] and precede progression to significant liver damage. Subtle chemical changes in body fluids can be assessed by a range of state-or-the-art analytical chemistry methods, including mass spectrometry or proton (^{1}H) nuclear magnetic resonance (NMR) spectroscopy [6]. In this chapter, the potential of the metabolic profiling using NMR spectroscopy techniques is illustrated from results of urinary and plasma or serum NMR studies of liver cirrhosis and the characterisation of hepatocellular carcinoma (HCC).

Bio-fluids Suitable for NMR Metabolic Profiling

Urine is a particularly useful body fluid to study because it is easy to collect, store and transport. As centres for metabolic profiling studies are not currently routinely available in hospitals, ease of transportation from participating sites, nationally and internationally, to the NMR laboratories is an important factor. There are disadvantages in analysing urine, for example the scope for contamination of the sample at collection and patients with advanced disease may be unable to produce a urine sample. Importantly the urinary proton NMR spectrum includes resonances from metabolites related to gut microbial metabolism as well as host cellular metabolism and drug metabolism.

Collection of a blood sample is often preferred to collection of a urine sample, since there is more control as to how and when the sample is obtained. However, the blood sample is generally analysed as plasma or serum, rather than as intact blood, and obtaining plasma or serum can be a cause of variability in the sample. For example, some additives to promote plasma separation have strong and interfering NMR peaks, such as ethylenediaminetetraacetic acid (EDTA), and so collection of plasma into heparin tubes is preferred. The temperature and time taken to clot for serum samples should be recorded and consistent across comparator sites. For all samples, early processing and prompt freezing, if stored, is essential.

Low molecular weight metabolites detected by ^{1}H NMR metabolic profiling are in the μM concentration range and upwards. The chemical shift of the metabolites is influenced by the ionic concentration, pH and temperature of data collection. Notwithstanding this variability, the many resonances in a urinary or plasma ^{1}H NMR spectrum can be assigned with reference to published and freely available databases, for example the Human Metabolome Database (HMDB) [7].

NMR Facilities are Expanding

Clinical *in vivo* magnetic resonance spectroscopy (MRS) techniques have been used to study hepatic metabolism in adults since the mid-1980s [8, 9]. At that time, *in vivo* hepatic phosphorus-31 magnetic resonance techniques were used to assess energy metabolism (by interpreting the resonances from nucleoside triphosphates and inorganic phosphate) and phospholipid turnover (reflected by the levels of phosphomonoesters and phosphodiesters) [10]. For example, a number of clinical MRS studies using *in vivo* hepatic phosphorus-31 MRS have shown that the phosphomonester peaks increased and the phosphodiester peaks decreased with increasing severity of liver disease, which has been interpreted as increased membrane turnover in the failing liver [11]. Clinical *in vivo* proton MRS techniques have primarily been used to assess the lipid content of liver [12], but other metabolites such as choline and phosphatidylcholine can be detected [13].

In parallel with the development of *in vivo* clinical hepatic MRS studies, analysis of body fluids have become increasingly sophisticated [5]. Urine and plasma or serum have generally been the body fluid studied, but bile, faecal water and tissue extracts have also been analysed in detail. As it has become routine to analyse samples in NMR spectroscopy systems of magnetic fields of at least 11.7T (500MHz) the sensitivity of the technique has substantially improved. It is now standard to use 600MHz (14.1T) NMR spectrometers for high-throughput routine NMR analysis, with widely accepted sample preparation and sequence acquisition protocols [14].

URINARY NMR STUDIES

Metabolites Detected

NMR is a non-selective technique, which has the advantage that NMR signals will be detected from all proton-containing metabolites which are mobile (so low molecular weight metabolites rather than proteins) and of sufficient concentration (in the µM range and upwards).

The main component, typically 95%, of urine is water. Its concentration depends on the recent fluid intake of the subject and so the water peak, whilst characteristic in its resonance features, cannot be used as an internal NMR standard. Since the water concentration is in the molar range, and the metabolites of interest are in the µM-mM range or below then it is also necessary to suppress the water signal using appropriate sequences in order to observe the otherwise obscured smaller metabolite resonances.

Aside from water, the next components of highest concentration in urine from healthy subjects are urea and sodium chloride [15]. Suppression of the water signal will cause distortion of the NMR urea signal, as the water and urea protons are in rapid chemical exchange. Sodium chloride is only indirectly detectable in an NMR spectrum via subtle differences in chemical shift patterns.

Once the water and urea signals are suppressed, a representative urinary NMR spectrum consists of signals from hundreds of metabolites (Fig. **1**), and it is at this stage that the strengths of NMR spectroscopy can be realised. All low molecular weight proton containing metabolites will be detected if present within the sensitivity range of the magnetic field strength used. The most dominant metabolite in normal urine, after water and urea suppression, is creatinine, and so the two peaks from the two proton chemical environments of creatinine usually dominate a proton NMR spectrum [16]. However, up to 200 or so of the 2000 endogenous different metabolites present can be routinely identified and quantified in an NMR spectrum of human urine. Peaks can be readily assigned to creatinine, hippurate, citrate, creatine, lactate, alanine, glycine, glucose, acetate, dimethylamine (DMA), trimethylamine (TMA), phenylacetylglutamine (PAG), trimethylamine-*N*-oxide (TMAO) on the basis of chemical shift patterns [16].

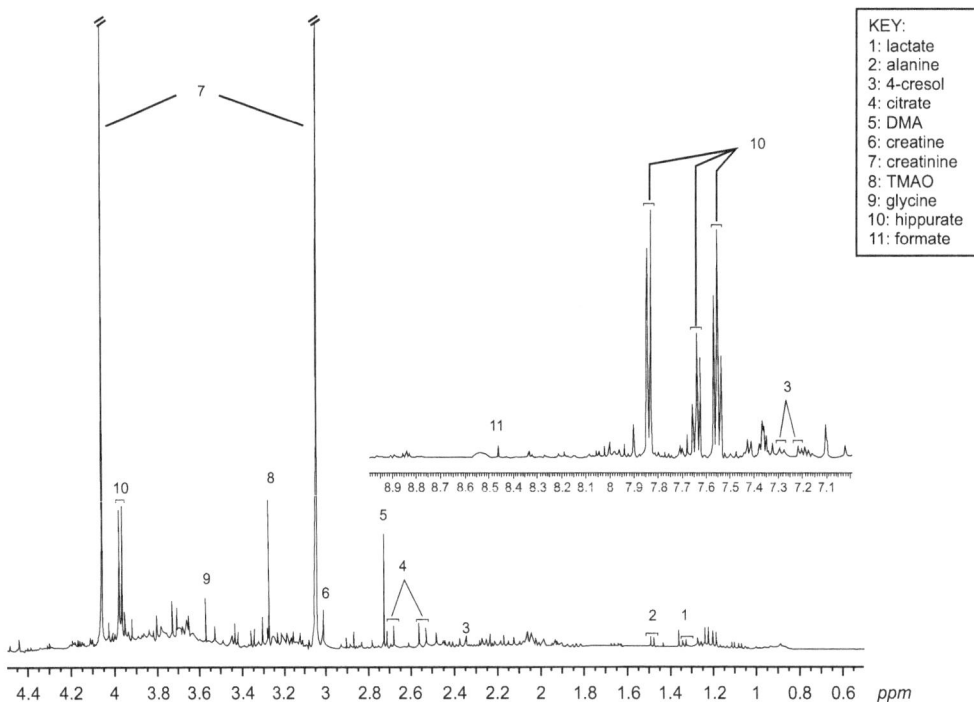

KEY:
1: lactate
2: alanine
3: 4-cresol
4: citrate
5: DMA
6: creatine
7: creatinine
8: TMAO
9: glycine
10: hippurate
11: formate

Fig. (1). An illustrative urinary proton NMR spectrum from a healthy male adult control subject, adapted from Williams *et al* [18].

Detecting other Factors of Urinary Chemical Composition

The chemical shift of many urine metabolite signals varies according to pH and ion composition, which limits the scope of automated identification of an NMR spectrum. If the source of the variation can be understood, this presents a source of additional information about urine composition. A strong relationship has been established between the concentration matrix, of both metabolites and inorganic ions, and the chemical shift matrix of metabolites [17]. This may allow for the concentrations of the NMR invisible inorganic ions, including Na^+, Cl^-, K^+, Li^+, Rb^+, Ca^{2+}, Mg^{2+}, Zn^{2+}, Al^{3+}, SO_4^{2-}, PO_4^{3-}, as well as albumin, to be indirectly determined from the chemical shift patterns specifically of creatinine, citrate and glycine.

Metabolites of Particular Interest

Urinary NMR spectral profiles are strongly influenced by differences in intestinal microbiota, as bacterial species may generate specific co-metabolites which are easily quantifiable by NMR analysis. Metabolites such as DMA, TMAO, hippurate, formate, PAG and 4-cresol sulphate have been implicated in gut microbial metabolism [18, 19], and NMR measurement of these may allow an assessment of the likely systemic consequences of differences in the gut microbiota. Since NMR spectroscopy is a non-selective technique, discerning which metabolites are endogenous from the host and which are microbiota-related can only be done by interpretation of the NMR spectral peaks and correlative analysis and not by the technique itself. The definition of a metabonomics technique specific for microbial metabolites is an important unmet need in this area. Human gut microbiota are dominated by the *Bacteroidetes* and *Firmicutes* phyla, and alterations in the ratio of *Firmicutes/Bacteriodetes* have been documented in a range of liver pathologies [20]. Attempts have been made to associate urine metabolites with operational taxonomic units [19]. For example a positive association between *Clostridia* spp. and hippurate levels has been noted [19]. Intestinal microbes influence formate metabolism, as it is a characteristic product in the mixed acid fermentation of the *Enterobacteriaceae* and has been extensively studied in *Escherichia coli*. Increased urinary formate could be consistent with increases in *Enterobacteriacea*, although there are other mechanisms for generating formate. 4-cresol sulfate has been recognized as the product of the bacterial metabolism of tyrosine and is only synthesized by few bacterial species, particularly by *Clostridia* spp. and, to a lesser extent, by *Bacteroidetes* [18]. PAG derives from phenylalanine, resulting from the bacterial putrefaction of proteins [21]. TMAO is produced through bacterial action on dietary carnitine and choline, and subsequent oxidation in the liver with flavin mono-oxygenases [22].

PLASMA AND SERUM NMR STUDIES

Proton NMR-based analysis of serum and plasma generates complex spectroscopic profiles, with resonances reflecting energy homeostasis in addition to lipid, protein and lipoprotein metabolism (Fig. **2**). The NMR profiles of sera and plasma are broadly similar, although some studies suggest there might be subtle differences in amino acid levels [23].

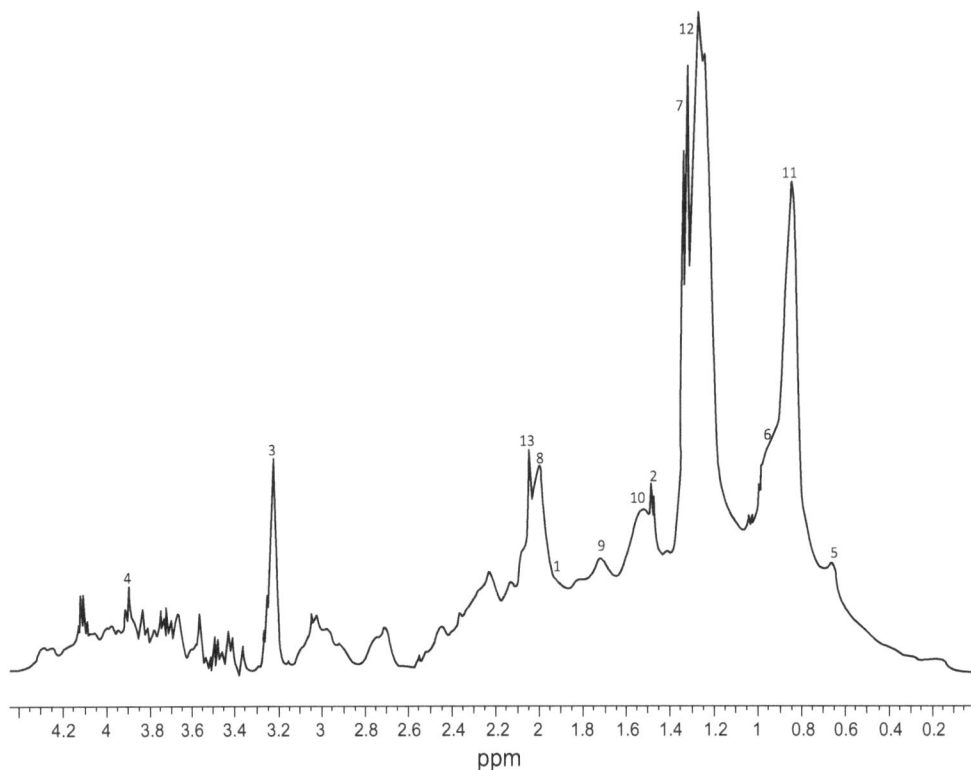

Fig. (2). An illustrative proton NMR spectrum of serum from a female healthy control (δ 0.00–4.50 shown). 1 Acetate, 2 alanine, 3 choline, 4 β-glucose, 5 HDL cholesterol, 6 isoleucine, 7 lactate, 8 lipid $CH_2C=C$, 9 lipid $CH_2CH_2C=C$, 10 lipid CH_2CH_2CO, 11 lipid mainly LDL, 12 lipid mainly VLDL, 13 N-acetylglycoprotein. Adapted from Williams *et al* [48].

Detailed multivariate data analysis allows interpretation of metabolic change, particularly once the NMR spectrum has been simplified by the use of a spin-echo sequence to filter out contributions from the lipoproteins (Fig. **3**).

A representative plasma NMR spectrum consists predominantly of signals from glucose and lipoproteins (including high-density lipoproteins (HDL), intermediate density lipoproteins (IDL), low-density lipoproteins (LDL), and very low-density lipoproteins (VLDL)) [24]. In addition, gut microbial metabolites are detected. Detection of branched chain amino acids is also straightforward.

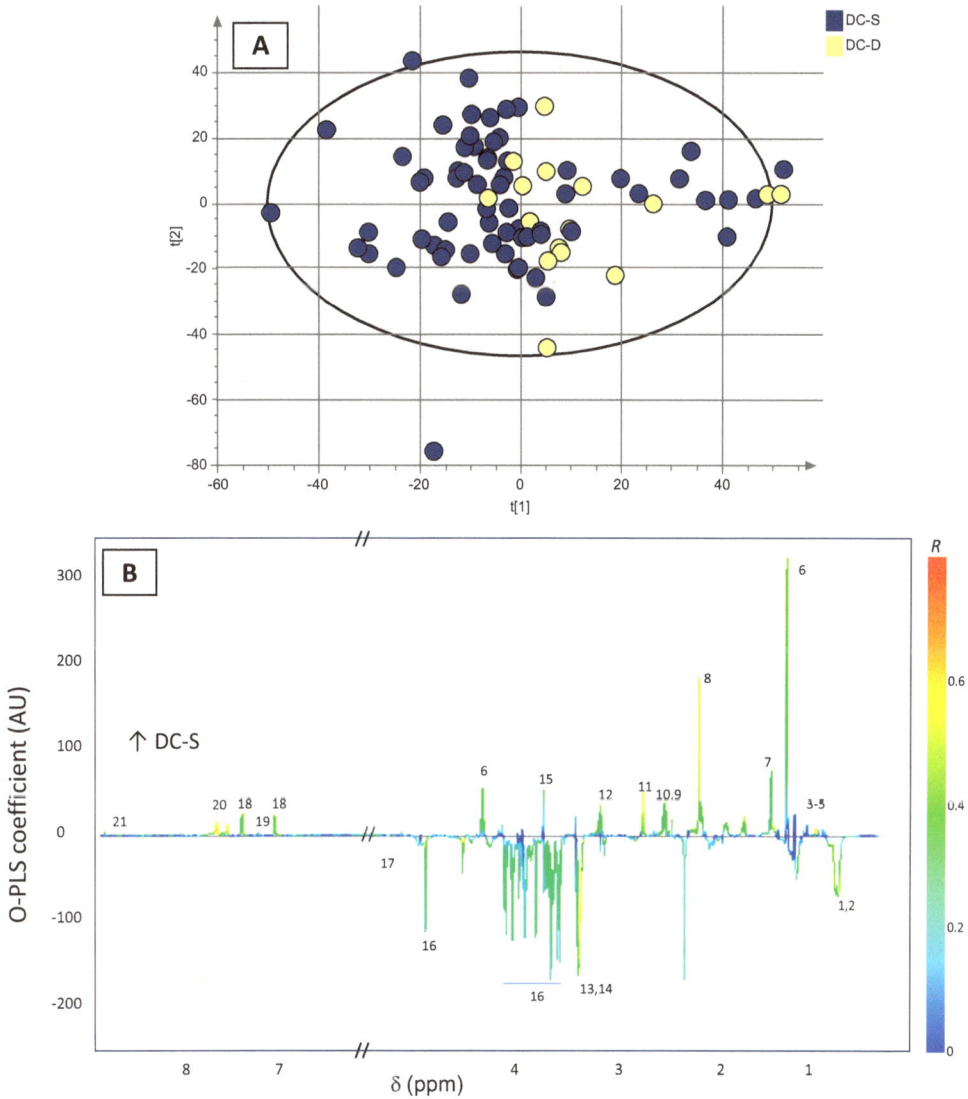

Fig. (3). Proton NMR plasma profile models. Multivariate models demonstrating discrimination of patients with decompensated cirrhosis who survive (DC-S) or die (DC-D) using the plasma NMR data. [A] Principal components analysis scores plot, three component model R2X = 0.75 Q2 = 0.54. [B] Orthogonal projection least squares S-line loading plot illustrating metabolites contributing to the discrimination of the two groups, including (6) lactate, (8) N-acetylated glycoproteins, (16) glucose and (18) tyrosine. Adapted from McPhail *et al* [26] under a creative commons attribution license https://creativecommons.org/licenses/by/4.0/.

The number of metabolites (expected and detected) in human serum has recently been recorded as reaching 25,000 [7], not all of which are detected by NMR.

Nevertheless, NMR studies of serum and plasma are particularly attractive and the metabolites detected reflect a number of important roles of the liver in glucose, lipid, amino acid and urea metabolism.

LIMITATIONS OF ANALYSING BODY FLUIDS

Confounding Factors

Factors that need to be considered routinely when interpreting a urinary, plasma or serum ^{1}H NMR spectrum include gender, diet, alcohol intake, co-morbidities and medication, including over-the-counter analgesic drugs such as paracetamol [25, 26]. Paracetamol metabolites and ethanol are readily identified in a urinary ^{1}H NMR spectrum, but without knowledge of a detailed dietary and medication history then interpretation of a ^{1}H NMR spectral change to a specific disease process may be misleading. Gut microbial co-metabolites add to the complexity of spectral interpretation [18].

Sample Collection

Blood plasma and serum analyses provide a 'snap-shot' assessment of metabolism. NMR data analyses of a blood sample from a single time point may not provide direct information about metabolic flux, and it is possible that metabolic challenge studies will be more helpful in assessing biochemical pathology [27].

Since urine may be stored in the bladder for some hours before the urine sample is collected, urine analysis tends to provide a summary assessment of metabolism over a short period of time.

Standardised NMR Acquisition Protocol

The method of sample preparation, method for water suppression, temperature of acquisition, and data acquisition parameters can all influence the appearance and results of the NMR spectrum. Standardised protocols have been described [14] and published NMR data sets can now be submitted to open access repositories [28].

Clinical Practice

NMR spectroscopy of urine and blood can be useful for characterising and assessing inborn errors of metabolism [29]. However, given the complexity of interpreting an NMR spectrum in the context of disease state and the range of confounding factors, then both urine and blood NMR are used as a research tool, with the emphasis on interpreting metabolite change in well-characterised clinical cohorts of patients.

EXAMPLES OF NMR PROFILING IN LIVER DISEASE

A number of studies have documented differences in urinary and plasma NMR profiles in a number of liver disorders, and illustrative metabolite differences are summarised in Table **1** and discussed in more detail below.

Table 1. Illustrative urine and plasma/serum NMR findings in cirrhosis, hepatic encephalopathy and hepatocellular carcinoma.

	NMR Findings		Interpretation	References
	Urinary NMR Biomarkers	**Plasma/Serum NMR Biomarkers**	**Microbiome *vs* Host**	
Cirrhosis *vs* controls	↓ acetate, alanine, glycine, hippurate, N-methyl nicotinic acid ↑ 1-methylnicotinamide	↓ lipids, choline, phosphocholine ↑ glucose, lactate, methionine, pyruvate	Host metabolism: decreased hepatic lactate clearance, decreased liver function, impaired lipid metabolism, alterations in muscle metabolism Microbiome co-metabolites: hippurate	[30, 31, 35]
Cirrhosis: overt HE *vs* no HE	↓ N-methyl nicotinic acid ↑ glutamate, histidine	↓ glucose, glycerol, lactate, methionine, TMAO ↑ alanine, branched chain amino acids, choline, glycine, lipids	Host metabolism: use of alternative energy sources, alterations in hepatic gluconeogenesis. Microbiome co-metabolites: a small effect	[31, 35]
HCC vs controls	↓ acetone, citrate, creatinine, DMA, hippurate, TMAO, ↑ acetate, carnitine, dimethylglycine	↓ lipid (mainly LDL) ↑ lipids (mainly VLDL), lactate, N-acetyl glycoproteins	Host metabolism: lipid turnover, markers of inflammation Microbiome co-metabolites: DMA, hippurate, TMAO	[40 - 42, 47, 49, 50]

Cirrhosis

The global incidence of cirrhosis is rapidly rising due to an increased prevalence of alcohol-related liver disease, obesity and viral hepatitis. Patients with cirrhosis are prone to episodes of decompensation, requiring hospital treatment and the prospect that the disease could progress to acute on chronic liver failure [30]. The current methods of outcome prediction in cirrhosis, for example the Model of End-Stage Liver Disease (MELD) score, Child-Pugh score or Lille model, have several limitations. In particular, they are not useful in patients with more systemic illnesses and the new chronic liver failure-sequential organ failure assessment (CLIF SOFA) score and its subdivisions are now the gold standard for phenotyping patients into those with stable disease, acute decompensation and acute on chronic liver failure. Patients with acute liver failure (ALF) have an unchanged definition and prognostication schema. Therefore, additional methods for predicting survival are sought, which might play a role in decision making for liver transplantation and resource allocation.

Plasma metabolic profiling has been investigated in minimal hepatic encephalopathy [31], cirrhotics with high MELD score [32], acute liver failure [33] and acute on chronic liver failure (ACLF) [34]. A serum fingerprint for ACLF was identified, in comparison chronic liver failure (CLF), with increased signals levels in ACLF from lactate, pyruvate, ketone bodies, glutamine, phenylalanine, tyrosine, and creatinine. High-density lipids were lower in ALCF than in the CLF group [34].

A recent study has examined whether survival can be accurately predicted in patients with decompensated cirrhosis [30]. Plasma metabolic profiling demonstrated highly accurate prognostication of acute decompensation, to the extent that plasma metabolic profiling was shown to be more accurate that standard clinically derived prognostic tools. The plasma NMR profiles of non-survivors were attributed to reduced choline/phosphocholine and lipid resonances, with increased lactate, tyrosine, methionine and phenylalanine signal intensities. Corroborating mass spectrometry studies confirmed that lysophosphatidylcholines (LPC) and phosphatidylcholines were downregulated in non-survivors, and it was hypothesised that circulating lipid abnormalities were a reflection of cell death. Larger data sets will be required to confirm the widespread clinical applicability of changes in LPC and amino acid dysregulation as a prognostic marker of mortality in decompensated liver cirrhosis [30].

Hepatic Encephalopathy

Hepatic encephalopathy (HE) is a common complication of cirrhosis, porto-systemic shunting of cirrhotics and a defining feature of acute liver failure [35]. The majority of patients with cirrhosis will experience an episode of HE at some point during their illness. This may manifest as confusion, somnolence, poor concentration or even coma. In some patients such overt symptoms are lacking, but when psychometric tests are performed, significant impairment is revealed in attention, concentration and executive function. Such covert HE (cHE) has been associated with progression to overt HE (OHE) and increased frequency of hospitalisation [36].

Hyperammonaemia is central to HE, and ammonia levels correlate with cerebral oedema and outcome of patients with ALF. However, in patients with cirrhosis, neither hyperammonaemia nor cerebral oedema correlate strongly with severity of HE and co-factors are required to explain the brain dysfunction. The role of the gut microbiota has recently been implicated [37], and has been explored in proof-of-principle studies assessing the effect of therapies, such as lactulose [38] and rifaximin [39].

A recent proton NMR serum study in patients with HE showed that supervised modelling provided discrimination between healthy controls and patients with cirrhosis [31]. A predictive model was generated which displayed strong discrimination between patients with and without cHE. cHE patients displayed increased serum concentrations of glucose, lactate, methionine, TMAO and glycerol, as well as decreased levels of choline, branch amino acids, alanine, glycine, acetoacetate, NAC, and lipid moieties. This remains to be validated in other centres.

If metabolic changes in urine samples of patients with cHE could be defined, then this would strengthen the validity of metabolic profiling as a diagnostic tool in patients with HE. In our recent study, the urinary ¹H-NMR metabolic profile from a well characterised group of patients with cirrhosis, in the presence and absence of HE, was interrogated with clinical and diagnostic assessments to determine the utility of urinary metabolic phenotyping for assessing cHE. Consistent with previous studies, we showed that urinary ¹H-NMR metabolic profiling discriminates patients with cirrhosis from healthy controls [35]. We identified a reduction in hippurate, alanine, acetate, glycine, and N-methyl nicotinic acid in cirrhotics compared to controls, with a concomitant an increase in 1-methylnicotinamide. We further demonstrated that urinary ¹H-NMR metabolic profiling has the potential to detect a urinary metabolic phenotype associated with OHE (albeit with less validity), but not cHE [35], in contrast to the previous plasma NMR study [31]. Expanding such studies to compare and contrast the NMR results from metabolic profiling of plasma and urine in a cohort of HE patients would be of value.

Hepatocellular Carcinoma

Significantly different proton urinary NMR profiles have been reported in HCC, compared to patients with cirrhosis and to healthy controls, across a range of metabolites, including reduced hippurate, citrate, DMA, TMAO and creatinine levels and increased acetate and carnitine levels, consistent with the diverse effects of liver cancer on metabolic pathways and interrelationship with the gut microbiome [40 - 42]. Findings in a Bangladeshi cohort [42] corroborated findings previous findings in Nigerian [40] and Egyptian [41] patient groups and the panel of metabolic markers identified could form the basis of a cost-effective

HCC dipstick screening test in low-resource countries.

The urinary metabolite changes observed in HCC subjects relate to alterations in both host and gut bacterial metabolism. A reduced urinary concentration of citrate, a tricarboxylic acid cycle intermediate, might be in agreement with Warburg's hypothesis of altered mitochondrial aerobic respiration and heightened physiological stress of cancer cells [41]. The alterations observed in carnitine may also be relevant; carnitine plays an essential role in mitochondrial metabolic pathways and can be absorbed from the diet or synthesis in the liver, kidney and testis. Carnitine is required for energy metabolism, enabling fatty acids to enter the mitochondria for β-oxidation [43]. Carnitine can be absorbed from the diet or synthesised in the liver, testis and kidney and reabsorbed via the renal system. Increased urinary carnitine will relate to excess carnitine ingestion, biosynthesis or poor reabsorption. With elevated metabolic activity and high cell-turnover, it has been hypothesised that elevated urine carnitine levels reflect tumour overproduction of carnitine to fuel mitochondrial activity and maintain rapid growth [41].

Advanced HCC can be complicated by cancer cachexia with associated sarcopenia, and urinary creatinine concentration has been suggested as a biomarker of sarcopenia [44]. In combination with other factors, urinary NMR creatinine concentration may contribute to a biomarker panel for HCC.

The urinary changes in DMA, hippurate and TMAO implicate alterations in the gut microbiome. TMAO is an aliphatic amine and the oxidation product of TMA, which is derived from dietary sources typically produced by bacterial degradation of dietary phosphatidylcholine and choline. A decrease in TMAO reflects dysregulation of the intestinal microbiota. DMA can also be a product of gut bacterial metabolism of dietary choline, although it can originate from the N-methylation of methylamines from the breakdown of creatine. Hippurate is an acyl glycine formed by the conjugation of benzoate with glycine in liver and kidney mitochondria. Benzoate is formed via the metabolism of gut microbes from dietary aromatic compounds via gut microbial metabolism. It is possible that the decreased hepatic function in patients with HCC (as evidenced by lower albumin and higher bilirubin levels) translates into less efficient benzoate conjugation and subsequently lower urinary hippurate excretion levels, and therefore hippurate levels may be considered as a surrogate marker of hepatic function [41].

Urinary NMR changes in HCC, consistent with the diverse effects of liver cancer on human physiology and gut bacterial action, may aid the development of a cost-effective HCC urinary dipstick screening test. Further validation work is required, but the pattern emerging from different studies around the world is promising.

Gut Microbial Modulation in Chronic Liver Disease

Fecal Microbiota Transplantation

On the basis of successful rescue therapies for *C. difficile* infection [45] and promising results in inflammatory bowel disease [46], the dysbiosis noted in patients with cirrhosis and HE has led to the development of therapies targeting gut microbial abnormalities.

A recent study in cirrhotic patients with recurrent HE, successfully treated by faecal microbiota transplantation (FMT), showed urinary metabolites were altered by antibiotics and FMT (reduced PAG, hippurate and formate) and returned to baseline post-FMT [21]. An open-label, randomized clinical trial with a 5-month follow-up in outpatient men with cirrhosis with recurrent HE on standard of care (SOC) was conducted with 1:1 randomization [21]. FMT-randomized patients received 5 days of broad-spectrum antibiotic pre-treatment, then a single FMT enema from the same donor with the optimal microbiota deficient in HE. The primary outcome was safety of FMT compared to SOC using FMT-related serious adverse events, and it was found that FMT with antibiotic pre-treatment was well tolerated. Cognition improved in the FMT group, but not the SOC, group and FMT increased diversity and beneficial taxa in the gut microbiota. The combined use of pre-treatment antibiotics and FMT makes it difficult to discern the precise role of FMT alone. However, this initial study is very encouraging and further studies are warranted to get greater insight into the gut microbial changes post-FMT in cirrhosis.

Liver Transplantation

Liver transplantation (LT) is life-saving in patients with advanced chronic liver disease and has multiple metabolic and immune benefits. Interestingly it has been associated with an amelioration of gut microbial composition. However, the effect of LT on microbial functionality, which can be related to overall patient benefit, is unclear and could affect the post-LT course. A recent study followed 40 outpatient patients with cirrhosis until 6 months after LT, investigating microbiota composition, microbiota functional analysis and cognitive tests [22]. There was an increase in urinary TMAO, which could be correlated with specific changes in serum lipids related to cell membrane products. The post-LT lipidomic profile appeared beneficial compared with the profile before LT. This study concluded that LT improves gut microbiota diversity and dysbiosis, which was accompanied by favourable changes in gut microbial functionality [22]. This study may be relevant as cardiovascular disease may occurs several years after LT, due to the effect of immunosuppressant medication. Longitudinal follow-up would be interesting to determine if increased TMAO levels persisted in some subjects and could be related to atherogenic events further along the post-LT course.

CONCLUDING REMARKS

NMR results from clinical studies in well documented groups of patients with liver disease illustrate the potential for using urinary and plasma or serum results to stratify groups of patients with liver disease, including HE and HCC, and to monitor treatment effects in cirrhosis [21, 22, 30, 35, 40, 41, 47]. Use of NMR metabolic profiling of body fluids to assess dysbiosis may prove to be particularly helpful in a clinical setting. In order to realise the potential for clinical metabolic profiling, multiple factors influencing the metabolic profile of the normal population need to be captured and considered in a lot of detail.

In summary, metabolic profiling studies of both urine and plasma or serum continue to expand. There is scope for using specific metabolite profiles as a simple and robust test for diagnosing and monitoring treatment effects for a range of disease phenotypes, including liver disease.

CONSENT FOR PUBLICATION

Not applicable.

CONFLICT OF INTEREST

The authors declare no conflict of interest, financial or otherwise.

ACKNOWLEDGEMENTS

We thank all our collaborators involved in the studies described, including Professor SD Taylor-Robinson, Imperial College London, Dr Jasmohan Bajaj, Gastroenterology, Hepatology and Nutrition, McGuire VA Medical Center and Virginia Commonwealth University, Richmond, Virginia, USA and Dr. Mamun-Al-Mahtab, Department of Hepatology, Bangabandhu Sheikh Mujib Medical University, Dhaka, Bangladesh.

GLOSSARY

^{1}H	proton
ACLF	acute on chronic liver failure
cHE	covert hepatic encephalopathy
CLF	chronic liver failure
DMA	dimethylamine

EDTA   ethylenediaminetetraacetic acid

FMT   faecal microbiota transplant

HCC   hepatocellular carcinoma

HE   hepatic encephalopathy

HMDB   Human Metabolome Database

LPC   lysophosphatidylcholine

MELD   Model of End-stage Liver Disease

MRI   magnetic resonance imaging

MRS   magnetic resonance spectroscopy

NMR   nuclear magnetic resonance

OHE   overt hepatic encephalopathy

SOC   standard of care

TMA   trimethylamine

TMAO   trimethylamine-*N*-oxide

PAG   phenylacetylglutamine

REFERENCES

[1]   Son, G.; Kremer, M.; Hines, I. N. Contribution of gut bacteria to liver pathobiology. *Gastroenterol Res Pract,* **2010**.

[2]   Tilg, H.; Cani, P.D.; Mayer, E.A. Gut microbiome and liver diseases. *Gut,* **2016**, *65*(12), 2035-2044. [http://dx.doi.org/10.1136/gutjnl-2016-312729] [PMID: 27802157]

[3]   Quigley, E.M.; Monsour, H.P. The gut microbiota and the liver: implications for clinical practice. *Expert Rev. Gastroenterol. Hepatol.,* **2013**, *7*(8), 723-732. [http://dx.doi.org/10.1586/17474124.2013.848167] [PMID: 24134195]

[4]   Williams, R.; Aspinall, R.; Bellis, M.; Camps-Walsh, G.; Cramp, M.; Dhawan, A.; Ferguson, J.; Forton, D.; Foster, G.; Gilmore, I.; Hickman, M.; Hudson, M.; Kelly, D.; Langford, A.; Lombard, M.; Longworth, L.; Martin, N.; Moriarty, K.; Newsome, P.; O'Grady, J.; Pryke, R.; Rutter, H.; Ryder, S.; Sheron, N.; Smith, T. Addressing liver disease in the UK: a blueprint for attaining excellence in health care and reducing premature mortality from lifestyle issues of excess consumption of alcohol, obesity, and viral hepatitis. *Lancet,* **2014**, *384*(9958), 1953-1997. [http://dx.doi.org/10.1016/S0140-6736(14)61838-9] [PMID: 25433429]

[5]   Cox, I.J.; Sharif, A.; Cobbold, J.F.; Thomas, H.C.; Taylor-Robinson, S.D. Current and future applications of *in vitro* magnetic resonance spectroscopy in hepatobiliary disease. *World J. Gastroenterol.,* **2006**, *12*(30), 4773-4783. [http://dx.doi.org/10.3748/wjg.v12.i30.4773] [PMID: 16937457]

[6]   Nicholson, J.K.; Holmes, E.; Kinross, J.M.; Darzi, A.W.; Takats, Z.; Lindon, J.C. Metabolic phenotyping in clinical and surgical environments. *Nature,* **2012**, *491*(7424), 384-392. [http://dx.doi.org/10.1038/nature11708] [PMID: 23151581]

[7]   Wishart, D.S.; Feunang, Y.D.; Marcu, A.; Guo, A.C.; Liang, K.; Vázquez-Fresno, R.; Sajed, T.; Johnson, D.; Li, C.; Karu, N.; Sayeeda, Z.; Lo, E.; Assempour, N.; Berjanskii, M.; Singhal, S.; Arndt, D.; Liang, Y.; Badran, H.; Grant, J.; Serra-Cayuela, A.; Liu, Y.; Mandal, R.; Neveu, V.; Pon, A.; Knox, C.; Wilson, M.; Manach, C.; Scalbert, A. HMDB 4.0: the human metabolome database for 2018. *Nucleic Acids Res.,* **2018**, *46*(D1), D608-D617.

[http://dx.doi.org/10.1093/nar/gkx1089] [PMID: 29140435]

[8]   Styles, P.; Scott, C.A.; Radda, G.K. A method for localizing high-resolution NMR spectra from human subjects. *Magn. Reson. Med.,* **1985**, *2*(4), 402-409.
[http://dx.doi.org/10.1002/mrm.1910020408] [PMID: 4094554]

[9]   Cox, I.J.; Bryant, D.J.; Ross, B.D.; Young, I.R.; Gadian, D.G.; Bydder, G.M.; Williams, S.R.; Busza, A.L.; Bates, T.E. Spectral resolution in clinical magnetic resonance spectroscopy. *Magn. Reson. Med.,* **1987**, *5*(2), 186-190.
[http://dx.doi.org/10.1002/mrm.1910050212] [PMID: 3657509]

[10]   Taylor-Robinson, S.D.; Sargentoni, J.; Bell, J.D.; Saeed, N.; Changani, K.K.; Davidson, B.R.; Rolles, K.; Burroughs, A.K.; Hodgson, H.J.; Foster, C.S.; Cox, I.J. In vivo and *in vitro* hepatic 31P magnetic resonance spectroscopy and electron microscopy of the cirrhotic liver. *Liver,* **1997**, *17*(4), 198-209.
[http://dx.doi.org/10.1111/j.1600-0676.1997.tb00806.x] [PMID: 9298490]

[11]   Bell, J.D.; Cox, I.J.; Sargentoni, J.; Peden, C.J.; Menon, D.K.; Foster, C.S.; Watanapa, P.; Iles, R.A.; Urenjak, J. A 31P and 1H-NMR investigation *in vitro* of normal and abnormal human liver. *Biochim. Biophys. Acta,* **1993**, *1225*(1), 71-77.
[http://dx.doi.org/10.1016/0925-4439(93)90124-J] [PMID: 8241291]

[12]   Szczepaniak, L.S.; Nurenberg, P.; Leonard, D.; Browning, J.D.; Reingold, J.S.; Grundy, S.; Hobbs, H.H.; Dobbins, R.L. Magnetic resonance spectroscopy to measure hepatic triglyceride content: prevalence of hepatic steatosis in the general population. *Am. J. Physiol. Endocrinol. Metab.,* **2005**, *288*(2), E462-E468.
[http://dx.doi.org/10.1152/ajpendo.00064.2004] [PMID: 15339742]

[13]   Cobbold, J.F.; Patel, J.H.; Goldin, R.D.; North, B.V.; Crossey, M.M.; Fitzpatrick, J.; Wylezinska, M.; Thomas, H.C.; Cox, I.J.; Taylor-Robinson, S.D. Hepatic lipid profiling in chronic hepatitis C: an *in vitro* and in vivo proton magnetic resonance spectroscopy study. *J. Hepatol.,* **2010**, *52*(1), 16-24.
[http://dx.doi.org/10.1016/j.jhep.2009.10.006] [PMID: 19913320]

[14]   Dona, A.C.; Jiménez, B.; Schäfer, H.; Humpfer, E.; Spraul, M.; Lewis, M.R.; Pearce, J.T.; Holmes, E.; Lindon, J.C.; Nicholson, J.K. Precision high-throughput proton NMR spectroscopy of human urine, serum, and plasma for large-scale metabolic phenotyping. *Anal. Chem.,* **2014**, *86*(19), 9887-9894.
[http://dx.doi.org/10.1021/ac5025039] [PMID: 25180432]

[15]   Mozolowski, W. Chemical composition of normal urine. *Lancet,* **1948**, *1*(6498), 423.
[http://dx.doi.org/10.1016/S0140-6736(48)90292-X] [PMID: 18904296]

[16]   Bouatra, S.; Aziat, F.; Mandal, R.; Guo, A.C.; Wilson, M.R.; Knox, C.; Bjorndahl, T.C.; Krishnamurthy, R.; Saleem, F.; Liu, P.; Dame, Z.T.; Poelzer, J.; Huynh, J.; Yallou, F.S.; Psychogios, N.; Dong, E.; Bogumil, R.; Roehring, C.; Wishart, D.S. The human urine metabolome. *PLoS One,* **2013**, *8*(9), e73076.
[http://dx.doi.org/10.1371/journal.pone.0073076] [PMID: 24023812]

[17]   Takis, P.G.; Schäfer, H.; Spraul, M.; Luchinat, C. Deconvoluting interrelationships between concentrations and chemical shifts in urine provides a powerful analysis tool. *Nat. Commun.,* **2017**, *8*(1), 1662.
[http://dx.doi.org/10.1038/s41467-017-01587-0] [PMID: 29162796]

[18]   Williams, H.R.; Cox, I.J.; Walker, D.G.; North, B.V.; Patel, V.M.; Marshall, S.E.; Jewell, D.P.; Ghosh, S.; Thomas, H.J.; Teare, J.P.; Jakobovits, S.; Zeki, S.; Welsh, K.I.; Taylor-Robinson, S.D.; Orchard, T.R. Characterization of inflammatory bowel disease with urinary metabolic profiling. *Am. J. Gastroenterol.,* **2009**, *104*(6), 1435-1444.
[http://dx.doi.org/10.1038/ajg.2009.175] [PMID: 19491857]

[19]   Li, M.; Wang, B.; Zhang, M.; Rantalainen, M.; Wang, S.; Zhou, H.; Zhang, Y.; Shen, J.; Pang, X.; Zhang, M.; Wei, H.; Chen, Y.; Lu, H.; Zuo, J.; Su, M.; Qiu, Y.; Jia, W.; Xiao, C.; Smith, L.M.; Yang, S.; Holmes, E.; Tang, H.; Zhao, G.; Nicholson, J.K.; Li, L.; Zhao, L. Symbiotic gut microbes modulate human metabolic phenotypes. *Proc. Natl. Acad. Sci. USA,* **2008**, *105*(6), 2117-2122.

[http://dx.doi.org/10.1073/pnas.0712038105] [PMID: 18252821]

[20]   Acharya, C.; Bajaj, J.S. Gut Microbiota and Complications of Liver Disease. *Gastroenterol. Clin. North Am.,* **2017**, *46*(1), 155-169.
[http://dx.doi.org/10.1016/j.gtc.2016.09.013] [PMID: 28164848]

[21]   Bajaj, J.S.; Kassam, Z.; Fagan, A.; Gavis, E.A.; Liu, E.; Cox, I.J.; Kheradman, R.; Heuman, D.; Wang, J.; Gurry, T.; Williams, R.; Sikaroodi, M.; Fuchs, M.; Alm, E.; John, B.; Thacker, L.R.; Riva, A.; Smith, M.; Taylor-Robinson, S.D.; Gillevet, P.M. Fecal microbiota transplant from a rational stool donor improves hepatic encephalopathy: A randomized clinical trial. *Hepatology,* **2017**, *66*(6), 1727-1738.
[http://dx.doi.org/10.1002/hep.29306] [PMID: 28586116]

[22]   Bajaj, J.S.; Kakiyama, G.; Cox, I.J.; Nittono, H.; Takei, H.; White, M.; Fagan, A.; Gavis, E.A.; Heuman, D.M.; Gilles, H.C.; Hylemon, P.; Taylor-Robinson, S.D.; Legido-Quigley, C.; Kim, M.; Xu, J.; Williams, R.; Sikaroodi, M.; Pandak, W.M.; Gillevet, P.M. Alterations in gut microbial function following liver transplant. *Liver Transpl.,* **2018**, *24*(6), 752-761.
[http://dx.doi.org/10.1002/lt.25046] [PMID: 29500907]

[23]   Suarez-Diez, M.; Adam, J.; Adamski, J.; Chasapi, S.A.; Luchinat, C.; Peters, A.; Prehn, C.; Santucci, C.; Spyridonidis, A.; Spyroulias, G.A.; Tenori, L.; Wang-Sattler, R.; Saccenti, E. Plasma and Serum Metabolite Association Networks: Comparability within and between Studies Using NMR and MS Profiling. *J. Proteome Res.,* **2017**, *16*(7), 2547-2559.
[http://dx.doi.org/10.1021/acs.jproteome.7b00106] [PMID: 28517934]

[24]   Aru, V.; Lam, C.; Kakimov, B.; Hoefsloot, H.C.J.; Zwanenburg, G.; Lind, M.V.; Schafer, J.; van Duynhoven, J.; Jacobs, D.M.; Smilde, S.K.; Engelsen, S.B. Quantification of lipoprotein profiles by nuclear magnetic resonance spectroscopy and multivariate data analysis. *Trends Analyt. Chem.,* **2017**, *94*, 210-219.
[http://dx.doi.org/10.1016/j.trac.2017.07.009]

[25]   Holmes, E.; Loo, R.L.; Stamler, J.; Bictash, M.; Yap, I.K.; Chan, Q.; Ebbels, T.; De Iorio, M.; Brown, I.J.; Veselkov, K.A.; Daviglus, M.L.; Kesteloot, H.; Ueshima, H.; Zhao, L.; Nicholson, J.K.; Elliott, P. Human metabolic phenotype diversity and its association with diet and blood pressure. *Nature,* **2008**, *453*(7193), 396-400.
[http://dx.doi.org/10.1038/nature06882] [PMID: 18425110]

[26]   Loo, R.L.; Coen, M.; Ebbels, T.; Cloarec, O.; Maibaum, E.; Bictash, M.; Yap, I.; Elliott, P.; Stamler, J.; Nicholson, J.K.; Holmes, E. Metabolic profiling and population screening of analgesic usage in nuclear magnetic resonance spectroscopy-based large-scale epidemiologic studies. *Anal. Chem.,* **2009**, *81*(13), 5119-5129.
[http://dx.doi.org/10.1021/ac900567e] [PMID: 19489597]

[27]   Dagnelie, P.C.; Menon, D.K.; Cox, I.J.; Bell, J.D.; Sargentoni, J.; Coutts, G.A.; Urenjak, J.; Iles, R.A. Effect of L-alanine infusion on 31P nuclear magnetic resonance spectra of normal human liver: towards biochemical pathology in vivo. *Clin. Sci. (Lond.),* **1992**, *83*(2), 183-190.
[http://dx.doi.org/10.1042/cs0830183] [PMID: 1327634]

[28]   Schober, D.; Jacob, D.; Wilson, M.; Cruz, J.A.; Marcu, A.; Grant, J.R.; Moing, A.; Deborde, C.; de Figueiredo, L.F.; Haug, K.; Rocca-Serra, P.; Easton, J.; Ebbels, T.M.D.; Hao, J.; Ludwig, C.; Günther, U.L.; Rosato, A.; Klein, M.S.; Lewis, I.A.; Luchinat, C.; Jones, A.R.; Grauslys, A.; Larralde, M.; Yokochi, M.; Kobayashi, N.; Porzel, A.; Griffin, J.L.; Viant, M.R.; Wishart, D.S.; Steinbeck, C.; Salek, R.M.; Neumann, S. nmrML: A Community Supported Open Data Standard for the Description, Storage, and Exchange of NMR Data. *Anal. Chem.,* **2018**, *90*(1), 649-656.
[http://dx.doi.org/10.1021/acs.analchem.7b02795] [PMID: 29035042]

[29]   Moolenaar, S.H.; Engelke, U.F.; Wevers, R.A. Proton nuclear magnetic resonance spectroscopy of body fluids in the field of inborn errors of metabolism. *Ann. Clin. Biochem.,* **2003**, *40*(Pt 1), 16-24.
[http://dx.doi.org/10.1258/000456303321016132] [PMID: 12542907]

[30]   McPhail, M.J.W.; Shawcross, D.L.; Lewis, M.R.; Coltart, I.; Want, E.J.; Antoniades, C.G.; Veselkov,

K.; Triantafyllou, E.; Patel, V.; Pop, O.; Gomez-Romero, M.; Kyriakides, M.; Zia, R.; Abeles, R.D.; Crossey, M.M.F.; Jassem, W.; O'Grady, J.; Heaton, N.; Auzinger, G.; Bernal, W.; Quaglia, A.; Coen, M.; Nicholson, J.K.; Wendon, J.A.; Holmes, E.; Taylor-Robinson, S.D. Multivariate metabotyping of plasma predicts survival in patients with decompensated cirrhosis. *J. Hepatol.,* **2016**, *64*(5), 1058-1067.
[http://dx.doi.org/10.1016/j.jhep.2016.01.003] [PMID: 26795831]

[31]   Jiménez, B.; Montoliu, C.; MacIntyre, D.A.; Serra, M.A.; Wassel, A.; Jover, M.; Romero-Gomez, M.; Rodrigo, J.M.; Pineda-Lucena, A.; Felipo, V. Serum metabolic signature of minimal hepatic encephalopathy by (1)H-nuclear magnetic resonance. *J. Proteome Res.,* **2010**, *9*(10), 5180-5187.
[http://dx.doi.org/10.1021/pr100486e] [PMID: 20690770]

[32]   Amathieu, R.; Nahon, P.; Triba, M.; Bouchemal, N.; Trinchet, J.C.; Beaugrand, M.; Dhonneur, G.; Le Moyec, L. Metabolomic approach by 1H NMR spectroscopy of serum for the assessment of chronic liver failure in patients with cirrhosis. *J. Proteome Res.,* **2011**, *10*(7), 3239-3245.
[http://dx.doi.org/10.1021/pr200265z] [PMID: 21568267]

[33]   Saxena, V.; Gupta, A.; Nagana Gowda, G.A.; Saxena, R.; Yachha, S.K.; Khetrapal, C.L. 1H NMR spectroscopy for the prediction of therapeutic outcome in patients with fulminant hepatic failure. *NMR Biomed.,* **2006**, *19*(5), 521-526.
[http://dx.doi.org/10.1002/nbm.1034] [PMID: 16598697]

[34]   Amathieu, R.; Triba, M.N.; Nahon, P.; Bouchemal, N.; Kamoun, W.; Haouache, H.; Trinchet, J.C.; Savarin, P.; Le Moyec, L.; Dhonneur, G. Serum 1H-NMR metabolomic fingerprints of acute-o--chronic liver failure in intensive care unit patients with alcoholic cirrhosis. *PLoS One,* **2014**, *9*(2), e89230.
[http://dx.doi.org/10.1371/journal.pone.0089230] [PMID: 24586615]

[35]   McPhail, M.J.; Montagnese, S.; Villanova, M.; El Hadi, H.; Amodio, P.; Crossey, M.M.; Williams, R.; Cox, I.J.; Taylor-Robinson, S.D. Urinary metabolic profiling by ^{1}H NMR spectroscopy in patients with cirrhosis may discriminate overt but not covert hepatic encephalopathy. *Metab. Brain Dis.,* **2017**, *32*(2), 331-341.
[http://dx.doi.org/10.1007/s11011-016-9904-0] [PMID: 27638475]

[36]   Weissenborn, K. The Clinical Relevance of Minimal Hepatic Encephalopathy--A Critical Look. *Dig. Dis.,* **2015**, *33*(4), 555-561.
[http://dx.doi.org/10.1159/000375348] [PMID: 26159273]

[37]   Bajaj, J.S. The role of microbiota in hepatic encephalopathy. *Gut Microbes,* **2014**, *5*(3), 397-403.
[http://dx.doi.org/10.4161/gmic.28684] [PMID: 24690956]

[38]   Bajaj, J.S.; Gillevet, P.M.; Patel, N.R.; Ahluwalia, V.; Ridlon, J.M.; Kettenmann, B.; Schubert, C.M.; Sikaroodi, M.; Heuman, D.M.; Crossey, M.M.; Bell, D.E.; Hylemon, P.B.; Fatouros, P.P.; Taylor-Robinson, S.D. A longitudinal systems biology analysis of lactulose withdrawal in hepatic encephalopathy. *Metab. Brain Dis.,* **2012**, *27*(2), 205-215.
[http://dx.doi.org/10.1007/s11011-012-9303-0] [PMID: 22527995]

[39]   Bajaj, J.S.; Heuman, D.M.; Sanyal, A.J.; Hylemon, P.B.; Sterling, R.K.; Stravitz, R.T.; Fuchs, M.; Ridlon, J.M.; Daita, K.; Monteith, P.; Noble, N.A.; White, M.B.; Fisher, A.; Sikaroodi, M.; Rangwala, H.; Gillevet, P.M. Modulation of the metabiome by rifaximin in patients with cirrhosis and minimal hepatic encephalopathy. *PLoS One,* **2013**, *8*(4), e60042.
[http://dx.doi.org/10.1371/journal.pone.0060042] [PMID: 23565181]

[40]   Shariff, M.I.; Ladep, N.G.; Cox, I.J.; Williams, H.R.; Okeke, E.; Malu, A.; Thillainayagam, A.V.; Crossey, M.M.; Khan, S.A.; Thomas, H.C.; Taylor-Robinson, S.D. Characterization of urinary biomarkers of hepatocellular carcinoma using magnetic resonance spectroscopy in a Nigerian population. *J. Proteome Res.,* **2010**, *9*(2), 1096-1103.
[http://dx.doi.org/10.1021/pr901058t] [PMID: 19968328]

[41]   Shariff, M.I.; Gomaa, A.I.; Cox, I.J.; Patel, M.; Williams, H.R.; Crossey, M.M.; Thillainayagam, A.V.; Thomas, H.C.; Waked, I.; Khan, S.A.; Taylor-Robinson, S.D. Urinary metabolic biomarkers of

hepatocellular carcinoma in an Egyptian population: a validation study. *J. Proteome Res.,* **2011**, *10*(4), 1828-1836.
[http://dx.doi.org/10.1021/pr101096f] [PMID: 21275434]

[42]   Cox, I.J.; Aliev, A.E.; Crossey, M.M.; Dawood, M.; Al-Mahtab, M.; Akbar, S.M.; Rahman, S.; Riva, A.; Williams, R.; Taylor-Robinson, S.D. Urinary nuclear magnetic resonance spectroscopy of a Bangladeshi cohort with hepatitis-B hepatocellular carcinoma: A biomarker corroboration study. *World J. Gastroenterol.,* **2016**, *22*(16), 4191-4200.
[http://dx.doi.org/10.3748/wjg.v22.i16.4191] [PMID: 27122669]

[43]   Vaz, F.M.; Wanders, R.J. Carnitine biosynthesis in mammals. *Biochem. J.,* **2002**, *361*(Pt 3), 417-429.
[http://dx.doi.org/10.1042/bj3610417] [PMID: 11802770]

[44]   Pahor, M.; Manini, T.; Cesari, M. Sarcopenia: clinical evaluation, biological markers and other evaluation tools. *J. Nutr. Health Aging,* **2009**, *13*(8), 724-728.
[http://dx.doi.org/10.1007/s12603-009-0204-9] [PMID: 19657557]

[45]   Kelly, C.R.; Khoruts, A.; Staley, C.; Sadowsky, M.J.; Abd, M.; Alani, M.; Bakow, B.; Curran, P.; McKenney, J.; Tisch, A.; Reinert, S.E.; Machan, J.T.; Brandt, L.J. Effect of Fecal Microbiota Transplantation on Recurrence in Multiply Recurrent Clostridium difficile Infection: A Randomized Trial. *Ann. Intern. Med.,* **2016**, *165*(9), 609-616.
[http://dx.doi.org/10.7326/M16-0271] [PMID: 27547925]

[46]   Moayyedi, P.; Surette, M. G.; Kim, P. T.; Libertucci, J.; Wolfe, M.; Onischi, C.; Armstrong, D.; Marshall, J. K.; Kassam, Z.; Reinisch, W.; Lee, C. H. Fecal microbiota transplantation induces remission in patients with active ulcerative colitis in a randomized controlled trial. *Gastroenterology,* **2015**, *149*(1), 102-109.

[47]   Shariff, M.I.; Kim, J.U.; Ladep, N.G.; Crossey, M.M.; Koomson, L.K.; Zabron, A.; Reeves, H.; Cramp, M.; Ryder, S.; Greer, S.; Cox, I.J.; Williams, R.; Holmes, E.; Nash, K.; Taylor-Robinson, S.D. Urinary Metabotyping of Hepatocellular Carcinoma in a UK Cohort Using Proton Nuclear Magnetic Resonance Spectroscopy. *J. Clin. Exp. Hepatol.,* **2016**, *6*(3), 186-194.
[http://dx.doi.org/10.1016/j.jceh.2016.03.003] [PMID: 27746614]

[48]   Williams, H.R.; Willsmore, J.D.; Cox, I.J.; Walker, D.G.; Cobbold, J.F.; Taylor-Robinson, S.D.; Orchard, T.R. Serum metabolic profiling in inflammatory bowel disease. *Dig. Dis. Sci.,* **2012**, *57*(8), 2157-2165.
[http://dx.doi.org/10.1007/s10620-012-2127-2] [PMID: 22488632]

[49]   Ladep, N.G.; Dona, A.C.; Lewis, M.R.; Crossey, M.M.; Lemoine, M.; Okeke, E.; Shimakawa, Y.; Duguru, M.; Njai, H.F.; Fye, H.K.; Taal, M.; Chetwood, J.; Kasstan, B.; Khan, S.A.; Garside, D.A.; Wijeyesekera, A.; Thillainayagam, A.V.; Banwat, E.; Thursz, M.R.; Nicholson, J.K.; Njie, R.; Holmes, E.; Taylor-Robinson, S.D. Discovery and validation of urinary metabotypes for the diagnosis of hepatocellular carcinoma in West Africans. *Hepatology,* **2014**, *60*(4), 1291-1301.
[http://dx.doi.org/10.1002/hep.27264] [PMID: 24923488]

[50]   Shariff, M.; Kim, J.U.; Ladep, N.G.; Gomaa, A.I.; Crossey, M.; Okeke, E.; Banwat, E.; Waked, I.; Cox, I.J.; Williams, R.; Holmes, E.; Taylor-Robinson, S.D.. The Plasma and Serum Metabotyping of Hepatocellular Carcinoma in a Nigerian and Egyptian Cohort using Proton Nuclear Magnetic Resonance Spectroscopy. *J Clin Exp Hepatol,* **2017**, *7*(2), 83-92.
[http://dx.doi.org/10.1016/j.jceh.2017.03.007] [PMID: 28663670]

CHAPTER 3

¹H NMR: A Powerful Tool for Lipid Digestion Research

Bárbara Nieva-Echevarría, Encarnación Goicoechea and **María D. Guillén**[*]

Food Technology, Faculty of Pharmacy, Lascaray Research Center, University of the Basque Country (UPV/EHU), Paseo de la Universidad nº 7, 01006 Vitoria-Gasteiz, Spain

Abstract: Recent studies on lipid digestion have demonstrated the suitability of one-dimensional Proton Nuclear Magnetic Resonance spectroscopy (¹H NMR) for simple, fast and accurate global study of the changes underwent by lipids under gastrointestinal digestive conditions. In the same run and without chemical modification of the sample, this technique allows the study not only of lipolysis reactions, but also that of the occurrence and extent of oxidation reactions taking place during digestion. By using spectral data and applying different approaches, ¹H NMR allows one the quantification of the various kinds of glyceryl structures and fatty acids in complex lipid mixtures. Moreover, this technique also provides valuable knowledge about the broad variety of primary and secondary oxidation products that may be generated from unsaturated lipids during digestion and could remain bioaccessible for intestinal absorption. Likewise, ¹H NMR facilitates the study of the bioaccessibility of certain minor lipidic compounds of interest, which is another important aspect of food digestion research that is difficult to tackle. This promising methodology overcomes many of the limitations of the techniques currently used for these purposes.

Keywords: Aldehydes, Bioaccessibility, Conjugated Dienes, Diglycerides, Fatty Acids, ¹H NMR, Hydroperoxides, Lipolysis, Lipid Oxidation, Monoglycerides.

1. INTRODUCTION

High resolution ¹H NMR has been successfully applied for research in the field of Food Science and Technology, especially in the case of food lipids [1 - 3]. Indeed, together with carbon and oxygen, hydrogen is one of the main elements present in all dietary fats and oils. Thus, from the ¹H NMR spectrum of any lipidic sample a great deal of information can be obtained in a simple, very fast and accurate way, and without any previous chemical modification of the sample.

---

[*] **Corresponding author María D. Guillén:** Food Technology, Faculty of Pharmacy, Lascaray Research Center, University of the Basque Country (UPV/EHU), Paseo de la Universidad nº 7, 01006 Vitoria-Gasteiz, Spain; Tel: 34-945-013081 ; E-mail: mariadolores.guillen@ehu.es

Atta-ur-Rahman & M. Iqbal Choudhary (Eds.)

As for the fundamentals of one-dimensional ¹H NMR technique, all the hydrogen atoms present in a molecule generate a signal in the spectrum. Depending on the local field of the proton (chemical environment), the generated peak will differ on the shape (multiplicity and coupling constant of the signal) and on the spectral region at which appears (chemical shift of the signal). In the case of protons showing the same chemical environment, signals at the same chemical shift will be generated. Thus, the study of the chemical shift, the multiplicity, the coupling constant and the relative intensity of the signals appearing in the ¹H NMR spectrum permit the identification of several functional groups containing protons present in the several molecules or groups of molecules that compose the sample. It must be noted that in comparison with ¹³C NMR spectra, in the case of ¹H NMR a more extensive overlapping of the spectral signals occurs because of the shorter range of chemical shifts. This fact obviously complicates the study of very complex lipid samples. Nevertheless, in this case, the use of more advanced NMR techniques has shown to be very helpful for structural elucidation and correct assignment of the spectral signals to the corresponding protons [4]. In addition, one of the main advantages of ¹H NMR is that the experimental time required to obtain valuable information is much shorter than in ¹³C NMR. This is due to the much shorter longitudinal relaxation time required for ¹H nuclei than for ¹³C nuclei [5].

Once the assignment of the signal to the corresponding proton/s is clear, by using spectral data it is possible to quantify accurately the several kinds of molecules supporting these protons that are present in the sample. For this purpose, the acquisition parameters used to record the spectrum must ensure full relaxation of the protons. Hence, the area of a spectral signal in the ¹H NMR spectrum is proportional to the number of protons that generate this signal, being the proportionality constant the same for all the spectral signals appearing in the spectrum [6]. Thus, to obtain relevant quantitative information about the molecules or group of molecules present in the sample, the simple integration of selected signals by means of the corresponding software and the development and application of specific equations are required.

Although classical methodologies based on long, multi-step, tedious and unspecific techniques still are widely employed, a large number of studies have demonstrated the usefulness of one-dimensional ¹H NMR in not only characterizing and quantifying major and minor lipidic components of foods, oils and fats, but also in studying their degradation process under different oxidative conditions, thus helping to shed light on the underlying mechanisms by which lipid degradation occurs [3, 7, 8]. It is well known that degradation processes affecting food lipids during technological processing and storage can be a major cause of food deterioration, generally with negative implications from the economic and health point of view. Furthermore, the nutritional quality and safety of lipids could also be modified during subsequent human gastrointestinal digestion. Since this physiological process is an inevitable step, it seems logical also to research the various chemical reactions that may affect food lipidic components under gastrointestinal digestive conditions, in order to better understand the effect of lipids on human health and also to be able to design healthier foods and diets. In this context, this chapter deals with the usefulness of applying ^{1}H NMR to study lipid digestion.

2. ^{1}H NMR A VERY USEFUL TOOL FOR THE STUDY OF HYDROLYSIS DURING FOOD LIPID DIGESTION

2.1. Lipolysis Reactions Under Gastrointestinal Conditions and Relevance of their Study

To date, the most studied of the chemical reactions occurring in the gastrointestinal tract and affecting lipids has been lipolysis. As is well known, food lipids are mainly made up of triglycerides (TG). However, these molecules cannot be directly absorbed in the gastrointestinal tract and hydrolytic cleavage of the ester bonds of TG is required before their incorporation into the organism. In recent years, primarily due to increasing public concern about obesity and related cardiovascular diseases, special attention has been paid to furthering knowledge about the factors affecting lipid bioaccessibility and bioavailability, in order to develop strategies that could modulate lipid uptake, and thus improve health status [9 - 12].

Lipolysis is catalyzed in the gastrointestinal tract by digestive lipases, which are present in saliva, gastric and duodenal juices. It must be noted that lingual lipase only plays an important role in lipid digestion during infancy [13]. In healthy adults, hydrolysis of TG mainly takes place in the duodenum through the action of pancreatic lipase (colipase-dependent lipase) at the lipid/water interface. When it comes to the hydrolysis of minor lipidic components, such as water-soluble esters, lipovitamins and phospholipids, the enzymes required are carboxyl ester hydrolase (known as bile salt dependent lipase) and phospholipase A2. The regiospecificity of digestive lipases determines the hydrolysis of TG, which proceeds in two steps. Fig. (**1**) shows the pathway generally described for lipolysis reaction during digestion [14 - 17]. Pancreatic lipase acts at the external ester bonds, releasing firstly a Fatty Acid (FA) molecule and transforming the original TG into one molecule of 1,2-diglycerides (1,2-DG). As diglycerides cannot be absorbed, the release of another FA molecule from the other external ester bond takes place. As a result, one molecule of 2-monoglyceride (2-MG) and two FA are generated.

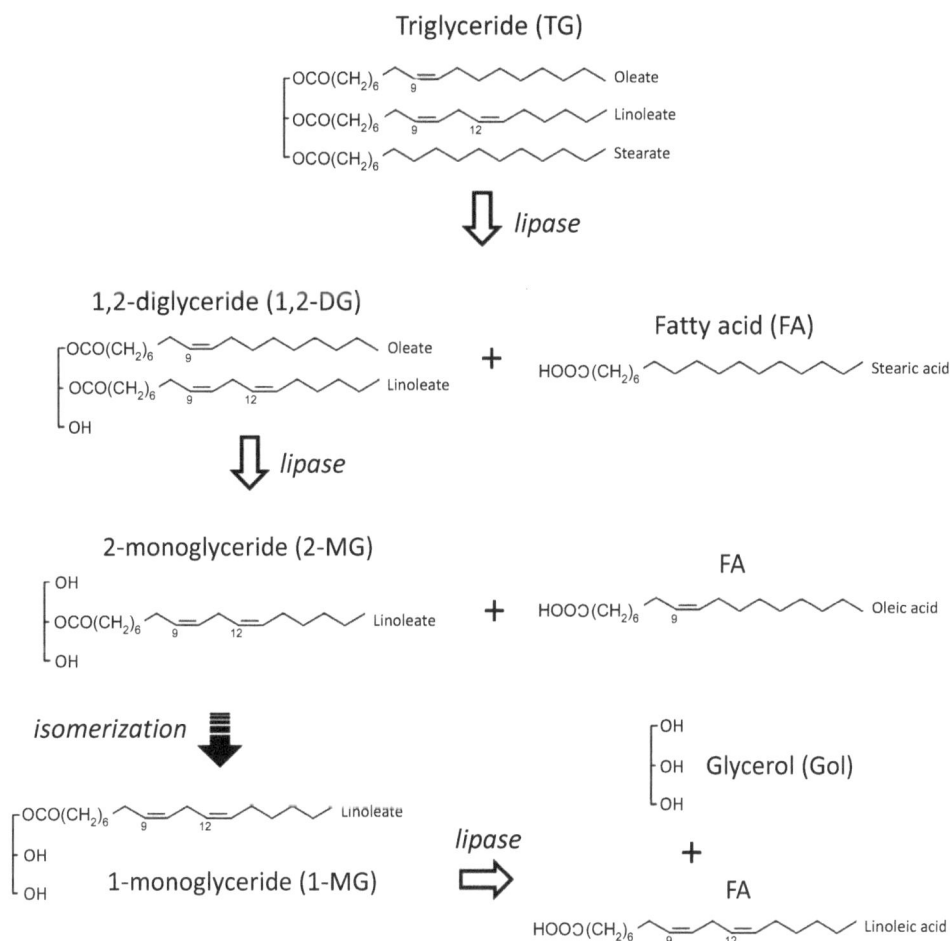

Fig. (1). Schematic representation of complete hydrolysis of triglycerides taking place during digestion [14 - 17].

It has also been described that, in addition to these molecular species, the occurrence of 1,3-diglycerides (1,3-DG) and 1-monoglycerides (1-MG) in a substantially lower proportion can also occur due to the migration in 1,2-DG of the acyl group (AG) in position 2 to position 3, and in 2-MG of the AG in position 2 to external position, respectively. Indeed, the occurrence of 1,3-DG in the micellar and oil phases of human intestinal aspirates in much lower concentration than 1,2-DG have been already reported [18]. Likewise, these molecular transformations would consequently allow a third hydrolysis reaction in the glyceride, yielding one molecule of Gol and a third FA. Pioneer studies using radio-labeled TG reported that 30-40% of those ingested were entirely split into

Gol and three FA in the lumen of small intestine, not only in rats [19], but also in humans [20, 21]. Furthermore, the activity of gastric lipase cannot be ruled out, bearing in mind that up to 30% of ester bonds might undergo hydrolysis in the stomach [13, 22 - 24]. This initial emulsification of dietary lipids in the stomach seems to be of great importance for the activity of pancreatic lipases in the subsequent step [25]. In short, it has been said that lipid digestion is a very efficient process, during which 90-95% of lipids contained in foods are absorbed in the form of MG and FA [26 - 28]. Once generated, MG and FA are solubilized in the intestinal lumen by bile salt micelles and unilamellar vesicles composed mainly of phospholipids and transported to the intestinal epithelium, where they will cross enterocyte membrane by either passive diffusion or active transport [9, 29].

2.2. Current Methodologies Employed to Study Lipolysis Reactions During Digestion

Most of the available literature in the field of lipid digestion research have employed chromatographic techniques to separate, identify and quantify DG, MG and FA generated during the hydrolysis of TG. The main ones are Thin Layer Chromatography coupled to Flame Ionisation Detector (TLC-FID) [30] or to video densitometry [23, 31 - 33]; High Performance Liquid Chromatography coupled to an Evaporative Light Scattering Detector (HPLC-ELSD) [13, 34, 35]; and Gas Chromatography equipped with FID (GC/FID) [36] or with a Mass Spectrometer (GC/MS) detector [37 - 39].

However, it must be taken into account that these are lengthy methodologies that require several preparation steps before analysis. These include lipid extraction, identification of the different kinds of lipolytic species by using standard compounds and the drawing of several calibration curves for the quantification of FA, MG, DG and TG, which is especially time-consuming. Moreover, in the case of the above-mentioned studies using GC/MS, the obtained lipid extract of the sample obtained must also be chemically modified to obtain two different fractions of Fatty Acid Methyl Esters (FAMEs) for analysis: one fraction coming from the esterification of FA and transesterification of AG present in TG, DG or MG in acid medium, and the other corresponding exclusively to transesterified AG obtained in alkaline medium by the sodium methoxide transesterification method [37 - 39]. These sample transformations not only lengthen the analytical task, but also favor the potential generation of artifacts that might affect quantitative determinations. A further disadvantage of the above-mentioned techniques is the large amounts of different polluting organic solvents needed, which make them not environmentally friendly.

Although simple and unspecific, another methodology commonly employed in the literature to study lipid digestion process is continuous titration of FA by means of pH-stat equipment. In this method, the volume of NaOH added to maintain the pH of the reaction medium at a constant value (above 7) is continuously recorded, which gives information about the FA molecules released over time [36, 39 - 44]. However, the information provided by this technique is very limited since the number of moles of TG that remains intact as well as the number of moles of DG and MG that could be generated during lipolysis remain unknown. Moreover, the accuracy of FA titration depends on several factors. These include the selection of the end point value of the pH and the ionization of FA present in the lipidic sample, the pH of the medium and its composition in bile salts, electrolytes and components with buffering capacity [39, 43, 45, 46].

Nevertheless, as will be described in the following sections, Proton Nuclear Magnetic Resonance (¹H NMR) can overcome the limitations of the methodologies currently employed to study lipolysis and thus can be considered a very advantageous alternative [47, 48].

2.3. Use of ¹H NMR to Study Qualitatively and Quantitatively Complex Lipid Mixtures of Glycerides and Fatty Acids

The usefulness of ¹H NMR in studying the lipid digestion process both qualitatively and quantitatively was first tested by using complex lipid mixtures made up of pure glycerides (TG, DG, MG) and FA [47]. The resulting ¹H NMR approach would later be used to study real digested lipid samples [48], as described in next section.

As a first step to study the above-mentioned complex lipid mixtures, the identification and assignment of the proton signals corresponding to each one of the lipolytic species that may co-exist in the gastrointestinal lumen was carried out, this is, TG, 1,2-DG, 1,3-DG, 2-MG, 1-MG and FA. For this purpose, several standard compounds of different unsaturation degrees and chain lengths (above 14 carbon atoms) were used. The ¹H NMR spectra of some of these standard compounds are shown in Fig. (**2**). The assignment of the various spectral signals to the corresponding protons of the different standard compounds is given in Table **1**. It must be mentioned that the ¹H NMR spectra were recorded on a Bruker Avance 400 spectrometer operating at 400 MHz and the acquisition parameters used were those employed in previous edible oil and fat studies [49 - 54], namely spectral width 6410 Hz, relaxation delay 3 s, number of scans 64, acquisition time 4.819 s and pulse width 90°.

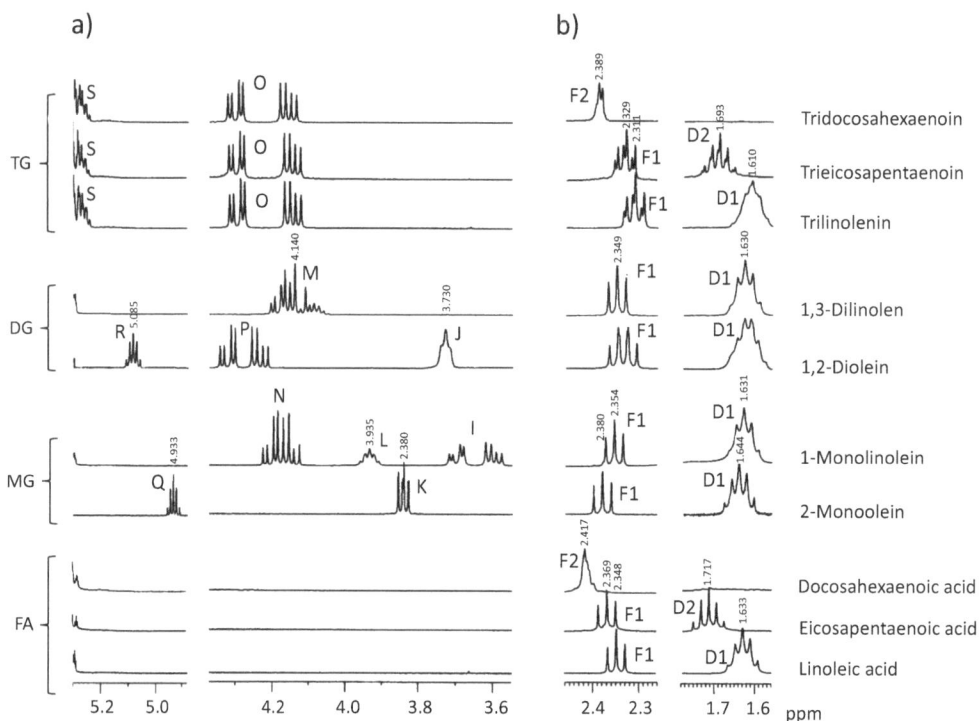

Fig. (2). Enlargement of several spectral regions where the most significant differences among the ¹H NMR spectra of tridocosahexaenoin (DHA C22:6ω-3), trieicosapentaenoin (EPA C20:5ω-3), trilinolenin (C18:3ω-3), 1,3-dilinolein (C18:2ω-6), 1,2-diolein (C18:1ω-9), 1-monolinolein, 2-monoolein, docosahexaenoic acid, eicosapentaenoic acid and linoleic acid can be noticed: **a)** signals of protons located in the glycerol backbone of TG, DG and MG, **b)** signals of protons located in *alpha-* and *beta-*position in relation to the carbonyl/carboxylic group. The signal letters agree with those in Table **1**.

Fig. (**2**) shows that the presence of TG, 1,2-DG, 1,3-DG, 1-MG or 2-MG in a lipidic sample can be rapidly detected by the occurrence of certain specific signals in the ¹H NMR spectral region from 3.5 to 5.3 ppm approximately. These signals are due to protons present in the glyceryl backbone of the above-mentioned glycerides, which differ not only in their chemical shifts, but also in their multiplicities (see Table 1). It may be observed in Fig. (**2a**) that protons in the glycerol backbone of TG generate two signals: signal **O** centered at 4.22 and signal **S** centered at 5.27 ppm. Those protons located in the glycerol backbone of 1,3-DG generate signal **M**, a multiplet ranging from 4.05 to 4.21 ppm, whereas the same kind of protons in 1,2-DG generate signal **J** at 3.73 ppm, signal **P** at 4.28 ppm and signal **R** centered at 5.08 ppm; in the case of 1-MG, three signals are also attributed to protons located in the glycerol backbone: signal **I**, **L** and **N** centered at 3.65, 3.94 and 4.18 ppm, respectively. In the case of 2-MG, the protons of the glycerol backbone generate signal **K** centered at 3.84 ppm and signal **Q** at 4.93 ppm. As could be expected, no proton signal associated to FA can be detected in this spectral region, because no glycerol backbone is contained in the molecule. Thus, as described above, protons located in the glycerol backbone of glycerides (either next to the ester bond or to the alcohol group) are greatly influenced by the chemical environment. In addition, this different chemical environment in glycerides also produces, to a limited extent, small differences in the chemical shift of signals due to protons bonded to carbon atoms in *alpha-* (signal **F**) and *beta-* (signal **D**) positions in relation to the carbonyl group of AG and to the carboxyl group of FA (see Fig. (**2b**) and Table **1**). As can be observed in Fig. (**2b**), only slight differences in the multiplicity and chemical shifts of signals **D1**, **D2**, **F1** and **F2** are noticed among AG supported on glycerides and FA. As far as protons bonded to carbon atoms in *gamma*-position and further in relation to the carbonyl group of AG and to the carboxyl group of FA are concerned, no differences can be observed in chains of the same nature, such as in, for example, linoleate/linoleic acid. In spite of this, simple observation of the ¹H NMR spectrum of a lipid sample can provide a great deal of information about the kind of the glyceride present (including the different isomers), because some of the signals generated by the protons present in these molecules can be properly identified.

Table 1. Chemical shift assignments and multiplicities of the ¹H NMR spectral signals in CDCl₃ of the main protons of triglycerides (TG), 1,2- and 1,3-diglycerides (1,2-DG and 1,3-DG), 2- and 1-monoglycerides (2-MG and 1-MG) and fatty acids (FA) of medium and long carbon atom chain [47, 48]. The signal letters agree with those given in Figs (2, 3, 5, 6 and 7).

Signal	Chemical Shift (ppm)	Multi-plicity	Functional Group	
			Type of Protons	**Compound**
A	0.88	t	-C**H**₃	saturated, monounsaturated ω-9 and/or ω-7 acyl groups and FA
	0.89	t	-C**H**₃	unsaturated ω-6 acyl groups and FA
B	0.97	t	-C**H**₃	unsaturated ω-3 acyl groups and FA
C	1.19-1.42	m*	-(C**H**₂)ₙ-	acyl groups and FA
D1	1.61	m	-OCO-CH₂-C**H**₂-	acyl groups in TG, except for DHA, EPA and ARA acyl groups
	1.62	m	-OCO-CH₂-C**H**₂-	acyl groups in 1,2-DG, except for DHA, EPA and ARA acyl groups
	1.63	m	-OCO-CH₂-C**H**₂-, COOH-CH₂-C**H**₂-	acyl groups in 1,3-DG, 1-MG and FA, except for DHA, EPA and ARA acyl groups
	1.64	m	-OCO-CH₂-C**H**₂-	acyl groups in 2-MG, except for DHA, EPA and ARA acyl groups

(Table 1) cont.....

Signal	Chemical Shift (ppm)	Multi-plicity	Functional Group	
			Type of Protons	**Compound**
D2	1.69	m	-OCO-CH$_2$-CH$_2$-	EPA and ARA acyl groups in TG
	1.72	m	COOH-CH$_2$-CH$_2$-	EPA and ARA acids
E	1.92-2.15	m**	-CH$_2$-CH=CH-	acyl groups and FA, except for -CH$_2$- of DHA acyl group in β-position in relation to carbonyl group
F1	2.26-2.36	dt	-OCO-CH$_2$-	acyl groups in TG, except for DHA acyl groups
	2.33	m	-OCO-CH$_2$-	acyl groups in 1,2-DG, except for DHA acyl groups
	2.35	t	-OCO-CH$_2$- COOH-CH$_2$-	acyl groups in 1,3-DG, 1-MG and FA, except for DHA acyl groups
	2.38	t	-OCO-CH$_2$-	acyl groups in 2-MG, except for DHA acyl groups
F2	2.37-2.41	m	-OCO-CH$_2$-CH$_2$-	DHA acyl groups in TG
	2.39-2.44	m	COOH-CH$_2$-CH$_2$-	DHA acid
G	2.77	t	=HC-CH$_2$-CH=	diunsaturated ω-6 acyl groups and FA
H	2.77-2.90	m	=HC-CH$_2$-CH=	polyunsaturated ω-6 and ω-3 acyl groups and FA
I	3.65	ddd	ROCH$_2$-CHOH-CH$_2$OH	glyceryl group in 1-MG
J	3.73	m***	ROCH$_2$-CH(OR')-CH$_2$OH	glyceryl group in 1,2-DG
K	3.84	m***	HOCH$_2$-CH(OR)-CH$_2$OH	glyceryl group in 2-MG
L	3.94	m	ROCH$_2$-CHOH-CH$_2$OH	glyceryl group in 1-MG
M	4.05-4.21	m	ROCH$_2$-CHOH-CH$_2$OR'	glyceryl group in 1,3-DG
N	4.18	ddd	ROCH$_2$-CHOH-CH$_2$OH	glyceryl group in 1-MG
O	4.22	dd,dd	ROCH$_2$-CH(OR')-CH$_2$OR''	glyceryl group in TG
P	4.28	ddd	ROCH$_2$-CH(OR')-CH$_2$OH	glyceryl group in 1,2-DG
Q	4.93	m	HOCH$_2$-CH(OR)-CH$_2$OH	glyceryl group in 2-MG
R	5.08	m	ROCH$_2$-CH(OR')-CH$_2$OH	glyceryl group in 1,2-DG
S	5.27	m	ROCH$_2$-CH(OR')-CH$_2$OR''	glyceryl group in TG
T	5.28-5.46	m	-CH=CH-	acyl groups and FA

Abbreviations: t: triplet; m: multiplet; DHA: docosahexaenoate; EPA: eicosapentaenoate; ARA: arachidonate(C20:4ω-6); d: doublet. *overlapping of multiplets of methylenic protons in the different acyl groups either in *beta*-position, or further, in relation to double bonds, or in *gamma*-position, or further, in relation to the carbonyl group; **overlapping of multiplets of the *alpha*-methylenic protons in relation to a single double bond of the different unsaturated acyl groups; ***this signal shows different multiplicity if the spectrum is acquired from the pure compound or taking part in the mixture.

Likewise, quantitative information about the molar proportions of these components can be obtained when considering the intensity of the above-mentioned non-overlapped signals and the number of protons that generate them. In ^{1}H NMR spectroscopy, if the relaxation delay and acquisition time allow the complete relaxation of the protons, the areas of the proton signals appearing in the spectrum are proportional to the number of protons that generate them and the proportionality constant is the same for all kinds of proton [6]. Based on this fact, several equations based on ^{1}H NMR spectral data were proposed in order to quantify the number of moles of the above-mentioned molecular species when present in complex lipid mixtures. It should be noted that for certain kind of glycerides different signals can be selected for integration, although the best option is to select those signals which are non-overlapped and generated by the highest number of protons, because the least error in their determination would take place.

Hence, the number of moles (N) of the different glycerides in a lipidic sample can be determined by using the following equations:

$$N_{2\text{-MG}} = Pc*A_Q = Pc*A_K/4 \qquad (1)$$

$$N_{1\text{-MG}} = Pc*A_L \qquad (2)$$

$$N_{1,2\text{-DG}} = Pc*A_R = (Pc*A_{3.54\text{-}3.75} -2*N_{1\text{-MG}})/2 \qquad (3)$$

$$N_{TG} = (Pc*2*A_{4.26\text{-}4.38} -2*N_{1,2\text{-DG}})/4 \qquad (4)$$

$$N_{1,3\text{-DG}} = (Pc*A_{4.04\text{-}4.38} -4*N_{TG} -2*N_{1\text{-MG}} -2*N_{1,2\text{-DG}})/5 \qquad (5)$$

where Pc is the proportionality constant existing between the area of the signals in a ^{1}H NMR spectrum and the number of protons that generate them (which could be easily determined by adding a known amount of an internal standard); A_Q, A_K, A_L and A_R are the area of the corresponding signals detailed in Fig. (**2**) and in Table 1; $A_{3.54\text{-}3.75}$ represents the areas of overlapped signals **I** and **J** ranging from 3.54 to 3.75 ppm; $A_{4.26\text{-}4.38}$ represents the area of the spectrum signals comprised between 4.26 and 4.38 ppm; $A_{4.04\text{-}4.38}$ represents the area of the spectrum signals **M**, **N**, **O** and **P** comprised between 4.04 and 4.38 ppm.

In spite of the potential partial overlapping of spectral signals, the determination of the number of moles of TG and 1,3-DG can be carried out by consecutive subtractions of the area of the signal generated by the number of moles of known glycerides (those previously determined using non-overlapped signals).

On the same basis, the number of moles of FA can be estimated from the area of the signals of the protons supported on carbon atoms in *alpha*-position in relation to the carbonyl and carboxyl groups of AG and FA respectively, by using the following equation:

$$N_{FA} = (Pc*A_{2.26-2.40}-6*N_{TG}-4*N_{1,2-DG}-4*N_{1,3-DG}-2*N_{1-MG}-2*N_{2-MG})/2 \qquad (6)$$

in which $A_{2.26-2.40}$ is the area of the spectrum signals appearing between 2.26 and 2.40 ppm. This equation is simplified considerably in the absence of 1,3-DG and 1-MG in digested lipid samples.

However, it must be pointed out that a different equation was proposed in the case of fish lipids containing docosahexaenoic (DHA, C22:6ω-3) FA/AG because of the slight overlapping existing between the signals of the protons supported on both carbon atoms in *alpha*- and *beta*-position in relation to the carboxyl/carbonyl group of DHA acid and AG (see signal **F2** in Fig. (**2b**)) and the signals of the protons supported on carbon atoms in *alpha*-position in relation to the carbonyl group of 1-MG, 2-MG and the carboxyl group of eicosapentaenoic acid (EPA, C20:5ω-3) (see signal **F1** in Fig. (**2b**)).

$$N_{FA} = (Pc*10A_{2.26-2.37}+Pc*5A_{2.37-2.44}-60*N_{TG}-40*N_{1,2-DG}-40*N_{1,3-DG}-18*N_{1-MG}-13*N_{2-MG})/20 \qquad (7)$$

where $A_{2.26-2.37}$ and $A_{2.37-2.44}$ are the areas of the spectrum signals between 2.26-2.37 and 2.37-2.44 ppm intervals, respectively.

It must be noted the area of signal **F** is due to methylenic protons bonded to carbon atoms in *alpha*-position in relation to carbonyl/carboxyl groups of any kind of compound present in the lipid sample; this includes AG and FA (either modified or not), as well as compounds formed during lipid oxidation process, such as aldehydes or ketones. However, as the oxidation level of the *in vitro* digested samples is estimated to be very low (as will be later discussed in subsections 3.3.1. and 3.3.2), the inclusion in this signal **F** of methylenic protons in *alpha*-position in relation to carbonyl/carboxyl groups different from those of AG or FA does not affect the above-mentioned calculations (equations in which A_F is considered), because the concentration of these other compounds is negligible in comparison with that of AG+FA. It must be noted that in the case of lipolyzed samples showing a very high oxidation level, the N_{FA} value hence calculated would be overestimated.

Once the total number of moles of TG and of any of the molecular species derived from them (N_T) is known, it is possible to determine the molar percentage of any of them (X) in a lipidic mixture:

$$X (\%) = 100*(N_X/N_T) \qquad \textbf{(8)}$$

$$N_T = N_{TG}+N_{1,2\text{-}DG}+N_{1,3\text{-}DG}+N_{2\text{-}MG}+N_{1\text{-}MG}+N_{FA} \qquad \textbf{(9)}$$

When data are expressed as molar percentages, there is no need to carry out the previous determination of the above-mentioned Pc because of its occurrence both in the numerator and denominator of equation 8. This fact considerably contributes to simplify quantitative determinations, being only necessary the areas of some spectral signals which are obtained by simply integration using the corresponding software.

To establish the accuracy of the proposed equations, their validation was carried out using 10 mixtures containing different known proportions of TG, 1,2-DG, 1,3-DG, 1-MG, 2-MG and FA [47]. The mixtures prepared covered a very broad range of concentrations and simulated edible oils and fats of vegetable and animal origins, including fish, with different hydrolysis levels. The molar percentages of the components of each of the mixtures were calculated by weight and also by the above-mentioned equations. Very low errors (between 0 and 9%) in the determination of the molar percentages were obtained, confirming the validity and great accuracy of the quantification performed by ¹H NMR. Unlike chromatographic methodologies, this approach provides a great deal of information in a simple and very fast way, without any previous chemical modification of the lipid sample, thus avoiding the formation of artifacts.

It should be noted that other authors previously employed ¹H NMR to quantify DG, MG or FA in fats and oils [53 - 61]. Nevertheless, these were only minor components and the estimated concentration was expressed in relation to TG. Equations allowing the accurate quantification by ¹H NMR of each of the molecular species that may arise from the hydrolysis of TG in a complex lipid mixture still remained to be developed.

2.4. Quantification of the Several Lipolytic Species and Assessment of the Extent of Lipolysis by ¹H NMR Spectroscopy in Real Digested Lipid Samples

The usefulness of the validated new methodology for the study of lipid digestion extent was afterwards proved by using real digested food lipid samples [48]. In this study, sunflower oil and sea bass meat were subjected to *in vitro* gastrointestinal digestion using different experimental conditions, in such a way that digested samples showing very different lipid composition and degree of hydrolysis were collected.

Before ^{1}H NMR study, extraction of the lipids of the *in vitro* digests is required. For this purpose, dichloromethane was used as solvent in a proportion of 2:3 (v/v) [47, 48, 62 - 67]. It should be noted that it is of paramount importance to ensure an exhaustive extraction for a proper study of lipolysis. Thus, after a first extraction, a second one is suitable after the acidification of the remaining water phase to pH≈2 with HCl. Thus the potential salts formed (if any) are dissociated and the FA present in the digest are protonated [36, 39]. Afterwards, both dichloromethane extracts are mixed and the solvent is eliminated by means of a rotary evaporator under reduced pressure at room temperature to avoid lipid oxidation.

As an example, Fig. (3) shows the ^{1}H NMR spectra of the lipid extracts of fish meat before digestion (not lipolyzed), partially and totally digested, together with some regions properly enlarged where the most significant changes can be detected. These are the following: i) from 1.55 to 1.75 ppm, where signals corresponding to methylenic protons bounded to carbon atoms in *beta*-position in relation to the carboxyl/carbonyl group of all FA/AG appear (except for DHA AG and FA); ii) from 2.25 to 2.45 ppm, where signals corresponding to methylenic protons bounded to carbon atom in *alpha*-position in relation to the carboxyl/carbonyl group of all FA/AG appear; and iii) from 3.5 to 5.2 ppm, where specific signals due to protons in the glycerol backbone of the different glycerides can be detected.

In the spectrum of non-digested fish lipid extract (see Fig. (3)), specific signals of the glyceryl backbone of TG (signals **O**, **S**) can be observed, together with the typical signals corresponding to the protons supported on either AG or FA (signals **A**, **B**, **C**, **D1**, **D2**, **E**, **F1**, **F2**, **G**, **H** and **T**). Furthermore, the absence of any other specific signal related to DG and MG points out that the main components of this lipid sample are TG, as could be expected. The occurrence of FA in this sample (if any) should be negligible. This is confirmed when checking the multiplicity and the chemical shift of signals **D1**, **D2**, **F1** and **F2**, which correspond to signals generated by protons located in TG.

By contrast, substantial changes are noticed in the ^{1}H NMR spectra of partially and totally lipolyzed fish lipid extracts (see below in Fig. (3)). It can be observed that as lipid digestion advances, signals **O** and **S** due to TG gradually disappear, whereas new signals appear and increase in intensity. These new signals are mainly those specific to 1,2-DG (signals **J**, **P** and **R**) and to 2-MG (signals **K** and **Q**). In addition, but at much lower intensity, there are also signals **I**, **L** and **N** due to protons in 1-MG (see the spectrum corresponding to totally lipolyzed fish lipid extract). The changes observed totally agree with the lipolysis reaction undergone under gastrointestinal conditions (see Fig. (1)). As previously mentioned, TG are consecutively transformed into 1,2-DG and this latter into 2-MG, due to the regiospecificity of pancreatic lipase. The occurrence of 1-MG is explained by the rearrangement of 2-MG into 1-MG by migration of the AG to external positions.

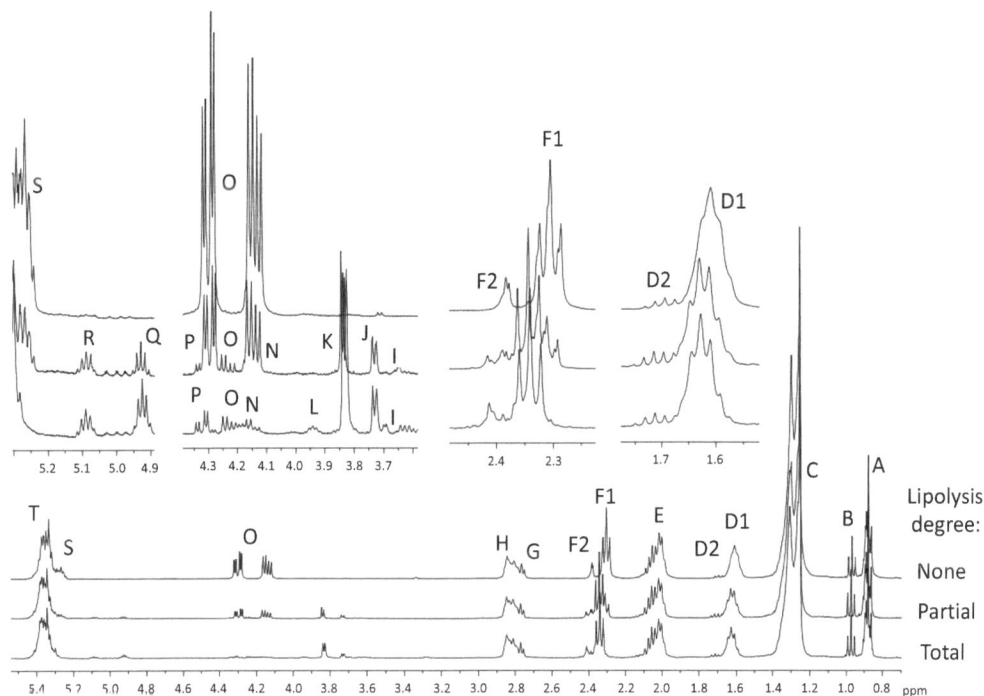

Fig. (3). ¹H NMR spectra of the lipid extracts of fish before digestion (not lipolyzed), during (partially lipolyzed) and after (totally lipolyzed) *in vitro* digestion process, together with enlargements of some spectral regions where the most significant changes are detected. The assignment of the signals is in agreement with Table **1**.

In addition to the above, other less noticeable changes also evidence the occurrence of lipolysis. For example, as the hydrolysis of TG proceeds, small differences in the chemical shift and multiplicity of signals **F1** and **F2** and of signals **D1** and **D2** can be found. Indeed, as shown in Fig. (**3**), these signals show higher chemical shifts in lipolyzed samples than in non-lipolyzed ones. Nevertheless, it must be pointed that the high degree of overlapping means that the simple observation of these proton signals cannot provide information about the proportion of 1,2-DG, 1,3-DG, 2-MG, 1-MG and FA generated.

Beside the qualitative description of the changes observed in the ¹H NMR spectra, the progression of lipid digestion was estimated applying the equations previously developed using mixtures of known composition made up with pure glycerides and FA of different nature [47]. The quantification of the different lipolytic species can be expressed in two different ways: i) as molar percentages of FA or AG supported in the different glyceryl backbone structures present (TG, DG, MG) in relation to the total number of AG plus FA present in the sample, as carried out in other studies focusing on the extent of lipid digestion [36, 39]; and ii) as the molar percentage of each glyceride (TG, 1,2- and 1,3-DG, 2- and 1-MG or Gol) in relation to the total number of glyceryl species regardless of their esterification degree.

2.4.1. Molar Percentages of Acyl Groups Bonded to the Several Kinds of Glycerides and of Fatty Acids in Relation to the Total Number of Acyl Groups Plus Fatty Acids

For the determination of lipolytic species in function of the total number of moles of AG plus FA (NT_{AG+FA}), the following equations were employed:

$$AG_{TG}\% = 100*(3*N_{TG})/NT_{AG+FA} \qquad (10)$$

$$AG_{1,2-DG}\% = 100*(2*N_{1,2-DG})/NT_{AG+FA} \qquad (11)$$

$$AG_{1,3-DG}\% = 100*(2*N_{1,3-DG})/NT_{AG+FA} \qquad (12)$$

$$AG_{2-MG}\% = 100*N_{2-MG}/NT_{AG+FA} \qquad (13)$$

$$AG_{1-MG}\% = 100*N_{1-MG}/NT_{AG+FA} \qquad (14)$$

$$FA\% = 100*N_{FA}/NT_{AG+FA} \qquad (15)$$

$$NT_{AG+FA} = 3*N_{TG}+2*N_{1,2-DG}+2*N_{1,3-DG}+N_{2-MG}+N_{1-MG}+N_{FA} \qquad (16)$$

where the meaning of the various variables are those described in equations [eq.1-eq.7].

Caution should be taken when using this quantification method because the molar percentages thus determined reflect only the FA and the AG bonded to the several kinds of glycerides. As a result, N_{TG} is multiplied by 3 in [eq.10] and $N_{1,2-DG}$ and $N_{1,3-DG}$ in [eq.11] and [eq.12], respectively. It must be noted that this way of quantifying does not reflect the relative proportions existing among the different kinds of lipolytic species considered as whole molecules [36].

2.4.2. Molar Percentages of the Several Kinds of Glycerides Including Glycerol in Relation to the Total Number of Glyceryl Species

In order to give another view of the extent of the lipolytic process at a molecular level, a second approach, based on the quantification in function of the total number of moles of glyceryl species (NT_{GS}), is more suitable. This type of quantification allows the application of mass balances because the proportion of the different lipolytic products totally reflects the stoichiometry of reaction [30, 68]. However, the determination of the number of moles of Gol is required (N_{Gol}). This is not extracted together with glycerides and FA because of its high polarity, and thus is absent in the lipid extract of digest. To this end, an indirect quantification of Gol based on the stoichiometry of the hydrolysis reaction is carried out.

$$N_{Gol} = (N_{FA}-N_{1,2-DG}-N_{1,3-DG}-2*N_{2-MG}-2*N_{1-MG})/3 \tag{17}$$

$$TG\% = 100*N_{TG}/NT_{GS} \tag{18}$$

$$1,2\text{-}DG\% = 100*N_{1,2-DG}/NT_{GS} \tag{19}$$

$$1,3\text{-}DG\% = 100*N_{1,3-DG}/NT_{GS} \tag{20}$$

$$2\text{-}MG\% = 100*N_{2-MG}/NT_{GS} \tag{21}$$

$$1\text{-}MG\% = 100*N_{1-MG}/NT_{GS} \tag{22}$$

$$Gol\% = 100*N_{Gol}/NT_{GS} \tag{23}$$

$$NT_{GS} = N_{TG} + N_{1,3-DG} + N_{1,2-DG} + N_{2-MG} + N_{1-MG} + NT_{Gol} \tag{24}$$

where the meaning of the various variables are the same before described in equations [eq.1-eq.5].

As was evidenced in subsections 2.4.1 and 2.4.2, this ¹H NMR methodology allows a global study of food lipolysis during digestion process, providing specific information at a molecular level that helps towards a better understanding of the hydrolysis reactions undergone. By using spectral data and taking into account that the area of a signal is proportional to the number of protons that generate it, lipolysis reaction can be characterized in detail in a fast and simple way, without any chemical modification and without the use of calibration curves of standard compounds.

2.4.3. Determination of the Extent of Lipid Digestion from ¹H NMR Spectral Data

This new approach based on ¹H NMR not only allows accurate quantification of the several lipolytic molecular species that may co-exist in the gastrointestinal lumen, such as TG, 1,2-DG, 1,3-DG, 2-MG, 1-MG, FA and Gol, but also the assessment of lipid digestion extent in terms of any of the different definitions employed to date (hydrolysis level, degree of TG transformation, lipid bioaccessibility level and percentage of FA physiologically releasable). These four parameters have been proposed in different studies carried out by several authors in an attempt to evaluate the level of lipid digestion in a single value. Nevertheless, it must be noted that the selection of one or another parameter is highly dependent on the methodology employed in the study to quantify lipolytic species:

- **Hydrolysis in the Chemical Sense (H_L%).** This parameter is one of the most commonly employed [30, 36, 39, 68], probably because it can be assessed using any of the methodologies before described in section 2.2. (including titration by means of pH-stat apparatus). It is defined by the percentage of FA released in relation to the total number of moles of AG plus FA. Thus, 100% is reached when all the ester bonds in any of the glycerides (TG, DG and MG) have been hydrolyzed. In the case of ^{1}H NMR, it can be determined by applying [eq.15].
- **Degree of Transformation of TG (T_{TG}%).** Instead of monitoring the evolution of the products generated, another approach to define the advance of a chemical reaction based on substrate disappearance [23, 32, 33]. If the initial sample is composed exclusively of TG, as is usually the case, this value can also be easily determined as follows:

$$T_{TG}\% = 100 - (N_{TG}/NT_{GS}) = 100 - (3*N_{TG}/NT_{AG+FA}) \tag{25}$$

where the variables of this equation have the same meaning as those previously mentioned in equations [eq.4, eq.16, eq.24].

However, it must be noted that the assessment of this parameter cannot be achieved employing pH-stat equipment and requires methodologies which are able to quantify the different kinds of glyceryl species (TG, DG and MG). Likewise, when using this parameter, a total lipolysis could be achieved if all the TG initially present in the sample were converted into DG and FA, in other words when a H_L value of only 30% is reached.

- **Lipid Bioaccessibility (L_{BA}%).** As digestion is a physiological process, some authors consider it more appropriate to introduce the notion of bioaccessibility when defining the extent of lipolysis reaction [30, 35]. In this case, attention focuses on the generation of lipidic species that can be absorbed in the intestinal tract, *i.e.* FA and MG. As with the parameter T_{TG}%, this parameter L_{BA}% can only be estimated if the methodologies employed allow the identification and quantification of the several molecular species arising from lipolysis reaction (TG, DG, MG and FA), such as the chromatographic ones. In the case of ^{1}H NMR spectroscopy, lipid bioaccessibility can be determined from the following equation that represents the sum of the molar percentages of AG bonded to 2- and 1-MG and that of FA in relation to the total number of moles of AG plus FA present:

$$L_{BA}\% = 100*(N_{1\text{-}MG}+N_{2\text{-}MG}+N_{FA})/NT_{AG+FA} \tag{26}$$

where the variables have the same meaning as previously described in equations [eq.1, eq.2, eq.6, eq.7, eq.16].

- **Percentage of FA Physiologically Releasable (FA_{PR}%).** In certain studies, which monitor lipid digestion by the continuous titration of FA with alkali, complete hydrolysis of TG into Gol and three FA is not contemplated and it is assumed that only two FA can be released from one TG molecule [24, 42, 44, 69, 70]. In these studies, a parameter defined as the percentage of FA physiologically releasable (FA_{PR}%) is proposed. This can also be determined by ¹H NMR spectral data by means of the following equation:

$$FA_{PR}\% = 100*N_{FA}/(2*N_{TGi}) \tag{27}$$

where the meaning of N_{FA} is the same as in equations [eq.6, eq.7] and N_{TGi} is the number of moles of TG initially present in the sample and could be estimated by using spectral data as follows:

$$N_{TGi} = NT_{AG+FA}/3 = NT_{GS} \tag{28}$$

where the meaning of the several variables are the same before described in equations [eq.1-7, eq.16, eq.17, eq.24].

In summary, it is possible from ¹H NMR data to determine all the parameters proposed to date by different research groups using other methodologies. This offers a significant benefit because it enables not only the comparison among different parameters in order to analyze their equivalence, if any [48], but also the comparison of results obtained by ¹H NMR with those reported in other studies in which different methodologies were used.

3. ¹H NMR A VERY USEFUL TOOL FOR THE STUDY OF LIPID OXIDATION DURING FOOD DIGESTION

3.1. Lipid Oxidation During Digestion: A Complex Reaction Recently Evidenced under these Physiological Conditions

In addition to lipolysis, the potential occurrence of other reactions like lipid oxidation during gastrointestinal digestion cannot be discarded. In fact, food lipids are unavoidably exposed to the conditions of the gastrointestinal environment, which have been described as pro-oxidant because of the occurrence not only of reactive species (metallic ions, oxygenated species) present in the food bolus or released from other food components like metalloproteins, but also of the presence of residual amounts of oxygen incorporated into food during mastication, as well as of hydrogen peroxide which can be secreted, all together combined with the low pH of gastric juice [71]. Furthermore, when lipids are hydrolyzed, their esterification degree decreases, thus increasing the proportion of FA in the medium. It is well known that FA are more prone to oxidative degradation than TG at the same unsaturation degree [72]; this is, the lower the esterification degree, the higher the susceptibility to oxidation (FA>MG>DG>TG).

Nearly two decades ago pioneer studies evidenced the occurrence of lipid oxidation during digestion and indicated the potential harmful effects derived [71, 73]. Since then, several studies have been performed using either *in vitro* digestion models or *in vivo* ones with animals, in order to deepen knowledge of the potential occurrence of oxidation during this physiological process [74 - 78].

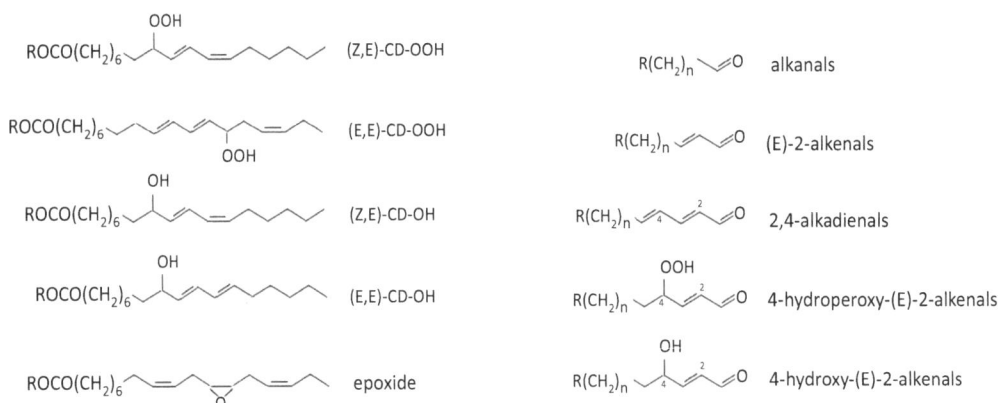

Fig. (4). Some of the possible oxidation compounds containing the functional groups that have been detected by ¹H NMR spectroscopy in *in vitro* digested lipid samples [62 - 67].

In short, according to the classical theoretical mechanism, during lipid oxidation unsaturated AG (and FA if present) can suffer many changes. During digestion, both hydrolysis and oxidation reactions could take place simultaneously. Firstly, the incorporation of oxygen to the unsaturated fatty acyl chains leads to the formation of unstable new compounds; these are named primary oxidation compounds and usually support a hydroperoxy group and conjugated dienes, showing either *Z,E* or *E,E* isomerism. These intermediate compounds can evolve to give so called secondary (or further) oxidation compounds containing other oxygenated groups, such as hydroxides, epoxides, ketones, or aldehydes. Some of these oxygenated chains can break down and give rise to volatile compounds of low molecular weight, and others can remain joined to the truncated triglyceride. Nevertheless, it must be noted that although this may be the mechanism as traditionally described, nowadays it is well known that lipid oxidation is a much more complex process that may evolve through different pathways and lead to the generation of different kinds of oxidation products depending on the oxidative conditions and on the nature of the lipids involved [8, 79 - 82]. Fig. (**4**) shows some oxidation compounds containing the functional groups that have been detected by ¹H NMR in edible oils submitted to *in vitro* gastrointestinal conditions.

3.2. Methodologies Employed to Study Lipid Oxidation During Digestion

To date classical methodologies have mostly been employed to study the oxidative status of digested samples. These include: determination of lipid hydroperoxides by iodometric titration (Peroxide Value), by ferrous ion oxidation in the presence of xylenol-orange (FOX-2 assay) or of thiocyanate (ferric thiocyanate method); measurement of Thiobarbituric Acid Reactive Substances (TBARS) by using the widely employed so-called TBARS test; and absorbance in certain ultraviolet visible regions for determining conjugated dienes [73, 74, 83 - 85]. Other studies have also employed indirect measurements, like the loss of "antioxidant" compounds in the digestion vessel during the process and the monitoring of oxygen uptake [76, 77]. Nevertheless, the limited accuracy and specificity of the above-mentioned techniques has been widely discussed in literature [79, 81]. Indeed, these techniques cannot provide information about the specific identity of the oxidation products generated from dietary lipids that could remain bioaccessible in the lumen of gastrointestinal tract. Thus, the use of more useful techniques is required to assess the lipid oxidative status of digests.

To overcome the limitations of these classical methodologies, recent studies have employed chromatographic techniques to detect and quantify some compounds considered as specific oxidation markers during digestion: malondialdehyde, hexanal, 4-hydroxy-2-hexenal and 4-hydroxy-2-nonenal [77, 86, 87]. Nonetheless, these techniques require the use of large amounts of solvents and are very laborious, because they involve extraction and derivatization steps. These chemically transform the sample subject of study; it is well known that this kind of procedures can lead to several experimental errors. In addition, it must be remembered that, as mentioned before, lipid oxidation is a very complex process that may evolve through different pathways and lead to the generation of different kinds of oxidation products depending on the oxidative conditions and on the nature of the lipids implied [79], as there may be very different oxidation markers in each case. The fact that lipid oxidation can also occur without the occurrence of the oxidation markers selected in the above-mentioned studies necessitates the use of additional approaches which can provide a simple fast and broad picture of the oxidation status by means of as many oxidation markers as possible is necessary [81].

As a result, other studies have employed two innovative techniques, like Solid Phase Microextraction (SPME) followed by GC/MS and ^{1}H NMR [62 - 67, 75, 88]. The great sensitivity of SPME-GC/MS provides very useful information on the specific nature of a broad range of volatile compounds of low molecular weight that may arise during the oxidation of polyunsaturated lipids, such as acids, alcohols, ketones, aldehydes, furans, lactones and hydrocarbons. In addition to its above-described usefulness in studying lipolysis, ^{1}H NMR also allows the simultaneous study of a broad variety of well-known functional groups typically present oxidation compounds, including not only conjugated dienes, hydroperoxides and aldehydes, but also epoxides and hydroxides, which is crucial to avoid reaching erroneous conclusions regarding the occurrence and extent of oxidation reactions affecting food lipids under digestive conditions.

3.3. Global Study by ^{1}H NMR of Lipid Oxidation During Digestion

Since the pioneer works of Grootveld and colleagues on the monitoring by ^{1}H NMR of the deterioration process of edible oils and fats submitted to episodes of thermal stressing [89, 90], several subsequent studies have confirmed the usefulness of this technique in providing a holistic view of food lipid oxidation quickly and without any chemical modification of the sample [8, 49 - 53, 91]. Recent studies have evidenced that this usefulness of ^{1}H NMR can also be applied to lipid digestion research, not only to shed light on the occurrence of lipolysis as described before, but also of lipid oxidation [62 - 67].

In one run ^{1}H NMR can provide qualitative and quantitative information: i) on the degradation of unsaturated FA and AG; ii) on the formation and bioaccessibility of derived oxidation compounds (primary, secondary or further); as well as iii) on the bioaccesibility of minor lipidic components present in food before digestion. In the following sections some examples will be given, which will explain how to obtain all this information about lipid oxidation during digestion from ^{1}H NMR spectral data.

3.3.1. Degradation of Unsaturated Fatty Acids/Acyl Groups During Lipid In Vitro Digestion

As it is generally known, lipid oxidation initially involves firstly the degradation of unsaturated acyl chains. Thus, if this occurs, it is expected that the original composition in the several kinds of FA/AG will be modified, *i.e.* decreased ratio unsaturated:saturated acyl chains. However, it must be noted that when the extent of these degradative reactions is relatively small (as could be expected under gastrointestinal digestive conditions) no significant modification can be noticed in the original lipid composition, because the changes are below or close to levels of experimental error.

Few studies have reported the evolution of AG and FA during lipid *in vitro* digestion. Calvo and colleagues [92] observed a significant diminution of *omega*-3 and *omega*-6 FA concentration after *in vitro* digestion of microencapsulated walnut oil; similar results were also reported by Kenmogne-Domguia and colleagues [77] during the *in vitro* digestion of fish oil-in-water emulsions. Nevertheless, the methodology employed in these studies was GC-FID, which involves chemical transformation of the lipid extract into FAMEs.

It is well known that ¹H NMR allows the determination of the molar percentages of the several kinds of AG present in the TG of lipids of vegetable and animal origins [8, 49 - 53, 91, 93]. Nevertheless, once digested, TG are no longer the main components of lipidic samples, because partial glycerides (DG, MG and Gol) and FA are released. Thus, in the case of hydrolyzed samples, instead of assessing the molar percentage of the several kinds of AG/FA in relation to the number of moles of TG (as previously done in non-hydrolyzed samples), the number of moles of other glycerides and FA must be taken into account. The equations proposed for digested lipid extracts to estimate by means of ¹H NMR spectral data the molar percentages of the several kinds of AG/FA (total unsaturated, saturated, *omega*-3, diunsaturated *omega*-6 and monounsaturated AG/FA) have been described in detail in previous studies addressing the *in vitro* gastrointestinal digestion of lipids from either vegetable [62, 63, 65] or animal origin [64, 66, 67]. In these latter studies, the oxidation extent reached in digested samples was so low that no appreciable qualitative changes were observed in the ¹H NMR spectra without further enlargement of the proton signals related to unsaturated FA/AG present. Thus, in order to investigate if the extent of oxidation reactions taking place under *in vitro* digestion conditions would be enough to involve the degradation of unsaturated AG /FA, a quantitative detailed study of the ¹H NMR spectra was needed.

In summary, quantification by means of ¹H NMR evidenced that during *in vitro* digestion of fresh oils, such as sunflower (rich in *omega*-6 lipids) and flaxseed oil (rich in *omega*-3 lipids), there was no statistically significant decrease in polyunsaturated AG/FA. Nevertheless, in the case of slightly oxidized sunflower and flaxseed oils, the proportion of unsaturated AG/FA degraded during digestion was greater, especially for the most unsaturated lipid samples (flaxseed oil ones).

These results highlighted that the extent of lipid oxidation under gastrointestinal conditions could be thought very low, although the initial lipid oxidative status and unsaturation degree (before digestion) influenced the advance of oxidation reactions [62, 63]. Likewise, a ^{1}H NMR quantitative approach was successfully employed in subsequent studies addressing the *in vitro* digestion of more complex food matrices, such as mixtures made up with oil and protein [65] and real food systems like minced fish meat [66, 67]. By using this methodology, it would be fair to state that the degradation of highly unsaturated AG/FA in the presence of proteins during *in vitro* digestion of slightly oxidized oils was notably reduced [65].

3.3.2. Generation of Lipid Oxidation Products During Oil In Vitro Digestion

It is possible by means of ^{1}H NMR spectroscopy to detect the presence of oxidation compounds of very different natures in the lipid sample in a fast and simple way, by verifying the occurrence of non-overlapped signals in certain spectral regions. This is the case of spectral regions: i) between 2.6-3.5 ppm, where proton signals due to epoxides appear [94, 95]; ii) between 5.9-6.7 ppm, where signals corresponding to protons of conjugated dienic systems supported on chains having also hydroperoxy or hydroxy groups are observable [95 - 99]; and iii) between 9.4-9.9 ppm, where aldehydic proton signals are visible [8, 50, 89, 90]. Table **2** shows the chemical shift assignments and multiplicities of the ^{1}H NMR spectral signals corresponding to some lipid oxidation compounds present in samples before and after *in vitro* digestion [49, 51, 90, 91, 96 - 102]. Therefore, simple observation of the spectra of digested lipids enables the simultaneous detection of a number of oxidation compounds whose natures are very different.

Table 2. Chemical shift assignments and multiplicities of the ^{1}H NMR spectral signals in CDCl₃ corresponding to some lipid oxidation compounds present in samples before and after *in vitro* digestion in agreement with previous studies [49, 51, 90, 91, 96, 97, 99 - 102]. The signal letters agree with those given in Figs (5 - 7).

Signal	Chemical Shift (ppm)	Multi-plicity	Functional Group	
			Type of Protons	**Compound**
			*Conjugated Dienic Systems*	
-	5.40	ddt		
-	5.64	dd		*Z,E*-conjugated double bonds
-	5.94	dd	-C**H**=C**H**-C**H**=C**H**-	associated with hydroxy group (OH) in octadeca-di/tri-enoic **AG** and **FA**
a	6.45	ddd		
-	5.47	ddm		
-	5.76	dtm		*E,E*-conjugated double bonds
-	6.06	ddtd	-C**H**=C**H**-C**H**=C**H**-	associated with hydroperoxy group (OOH) in octadeca-di/tri-enoic **AG** and **FA**
b	6.27	ddm		

(Table 2) cont.....

Signal	Chemical Shift (ppm)	Multi-plicity	Functional Group	
			Type of Protons	Compound
-	5.51	dtm		
-	5.56	ddm	-C**H**=C**H**-C**H**=C**H**-	*Z,E*-conjugated double bonds associated with hydroperoxy group (OOH) in octadeca-di/tri-enoic **AG** and **FA**
-	6.00	ddtd		
c	6.58	dddd		
-	5.58	dd		
-	5.71	dd	-C**H**=C**H**-C**H**=C**H**-	*E,E*-conjugated double bonds associated with hydroxy group (OH) in octadeca-di/tri-enoic **AG** and **FA**
-	6.03	dd		
d	6.18	dd		
Bis-allylic protons in oxygenated octadecatrienoic **AG** *and* **FA**				
a*	2.93	t	=HC-C**H₂**-CH=C--CH=	9-hydroxy-10*E*,12*Z*,15*Z*- and 16-hydroxy-9*Z*,12*Z*,14*E*-octadecatrienoic **AG** and **FA**
Epoxides				
e	2.94	m	-C**H**O**H**C-	monoepoxy-octadecadienoic **AG** and **FA**
f	3.05	m	-C**H**O**H**C-	Other compounds containing epoxy-group?
g	3.20	m	-C**H**O**H**C-	Other compounds containing epoxy-group?
h	3.28	m	-C**H**O**H**C-	Other compounds containing epoxy-group?
Aldehydes				
i	9.56	d	-C**H**O	4-hydroxy-*E*-2-alkenals
j	9.57	d	-C**H**O	4-hydroperoxy *E*-2-alkenals
k	9.60	d	-C**H**O	*Z,E*-2,4-alkadienals
l	9.75	t	-C**H**O	alkanals

Abbreviations: d: doublet; t: triplet; m: multiplet; DHA: docosahexaenoate; AG: acyl group; FA: fatty acid.
*Only due to certain isomers of hydroperoxy- and hydroxy-octadecatrienoic AG/FA supporting also conjugated dienes.

Example 1: Effect of Initial Oxidative Status of Lipids on the Oxidation Reached During In Vitro Gastrointestinal Digestion of Sunflower Oil

As a first example, Fig. (**5**) shows the ¹H NMR spectra of fresh (Sun) and slightly oxidized (SunOx) sunflower oil, together with the lipid extracts obtained after their *in vitro* digestion (DigSun, DigSunOx) [62]. As it is expected, fresh sunflower oil before digestion (Sun spectrum) does not contain oxidation compounds. By contrast, after digestion (DigSun spectrum), signals **c** related to protons of *Z,E*-conjugated dienes supported on chains having also hydroperoxy groups (*Z,E*-CD-OOH) appear, highlighting the generation of this kind of primary oxidation compounds during *in vitro* digestion process of this initially non-oxidized oil. In the case of slightly oxidized sunflower oil samples, before digestion signals **b** and **c** due to *E,E*- and *Z,E*-conjugated dienic systems associated with hydroperoxy groups, respectively, can already be detected (SunOx spectrum in Fig. (**5**)). After *in vitro* digestion of slightly oxidized sunflower oil (DigSunOx spectrum in Fig. (**5**)), together with signals **b** and **c**, other new signals appear named signal **a**, due to *Z,E*-conjugated dienic systems supported on chains having also hydroxy groups (*Z,E*-CD-OH), and signal **d**. This latter signal was attributed to olefinic protons located in *E,E*-conjugated dienic systems associated with hydroxy groups (*E,E*-CD-OH).

To the best of our knowledge, this was the first time that the generation of hydroxy-conjugated dienes during lipid *in vitro* digestion was reported. Indeed, in previous *in vitro* digestion studies only the generation of lipid hydroperoxides was described [73, 74, 76, 84], because the methodologies employed to assess lipid oxidation level were not able to determine hydroxy-conjugated dienes. This new finding was made possible because ^{1}H NMR provides one-run simultaneous information on the occurrence of both hydroperoxy- and hydroxy-conjugated dienes (if present), because some of their proton signals are not overlapped.

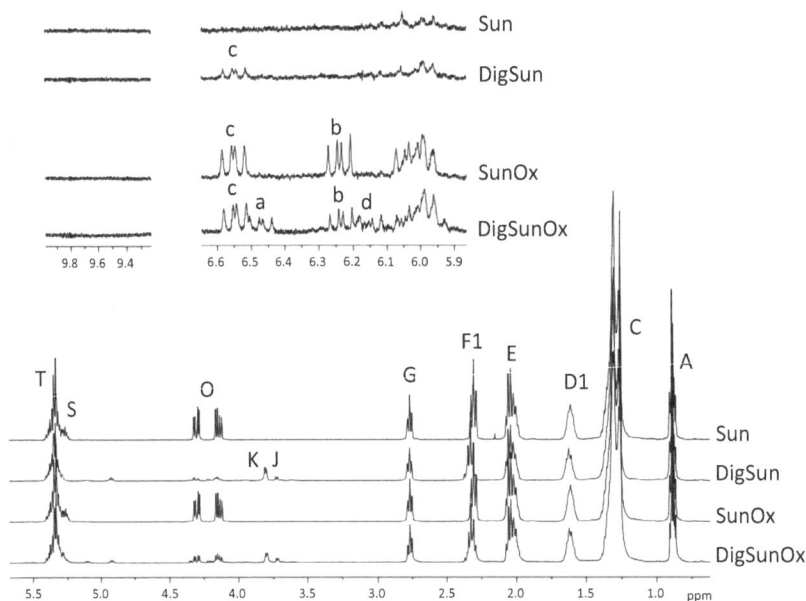

Fig. (5). ^{1}H NMR spectra of fresh and slightly oxidized sunflower oils (Sun, SunOx) together with the lipid extracts obtained after their *in vitro* digestion (DigSun, DigSunOx). Some spectral regions, where the most significant changes can be detected, are enlarged. The assignment of the signals is in agreement with Tables 1 and 2.

As far as the quantitative determination of primary oxidation products in digested lipid samples from ^{1}H NMR spectral data is concerned, several equations have been proposed to estimate the proximate concentration of *Z,E*- and *E,E*- isomers of CD-OOH and of CD-OH, expressed as mmol/mol of AG plus FA present [62, 63]:

$$E,E\text{-CD-OH (mmol/molAG+FA)} = 1000*(Pc*A_{6.18})/NT_{AG+FA} \qquad \textbf{(29)}$$

$$E,E\text{-CD-OOH (mmol/molAG+FA)} = 1000*(Pc*A_{6.27})/NT_{AG+FA} \qquad \textbf{(30)}$$

$$Z,E\text{-CD-OH (mmol/molAG+FA)} = 1000*(Pc*A_{6.45})/NT_{AG+FA} \qquad \textbf{(31)}$$

$$Z,E\text{-CD-OOH (mmol/molAG+FA)} = 1000*(Pc*A_{6.58})/NT_{AG+FA} \qquad \textbf{(32)}$$

where $A_{6.18}$ is the area of signal **d** at 6.18 ppm corresponding to one proton of the *E,E*-conjugated double bond supported on chains having also a hydroxy group; $A_{6.27}$ is the area of signal **b** at 6.27 ppm due to one proton of the *E,E*-conjugated double bond supported in chains having also a hydroperoxy group; $A_{6.45}$ is the area of signal a at 6.45 ppm corresponding to one proton of the *Z,E*-conjugated double bond supported on chains having also a hydroxy group; $A_{6.58}$ is the area of signal **c** at 6.58 ppm due to one proton of the *Z,E*-conjugated double bond supported in chains having also a hydroperoxy group, (see Table **2**) and NT_{AG+FA} can be obtained using equation [eq.16].

As previously commented in subsection 2.3, it should be pointed out that using equation [eq.16] NT_{AG+FA} might be overestimated when the lipolyzed sample shows a certain oxidation level because, in addition to intact AG and FA, some other compounds containing methylenic protons bonded to carbon atoms in *alpha*-position in relation to carbonyl/carboxyl group (modified AG or FA, aldehydes, ketones...) also contribute to the area of signal **F**. Anyway, in such cases, the concentration of the lipid oxidation products determined in digested samples would be higher than the approximate value obtained after applying equations [eq.29-32].

Example 2: Effect of Lipid Unsaturation Degree on the Oxidation Reached During In Vitro Gastrointestinal Digestion. Study of In Vitro Digestion Of Flaxseed Oil

Another similar study to that carried out with sunflower oil, intended as a model of *omega*-6 rich lipids [62], was also performed using flaxseed oil, selected as model of *omega*-3 rich lipids [63]. Fig. (**6**) shows the ¹H NMR spectra of samples similar to those of Fig. (**5**), but in this case of flaxseed oil: fresh and slightly oxidized flaxseed oil (Flax, FlaxOx) together with the lipid extracts obtained after their *in vitro* digestion (DigFlax, DigFlaxOx). Before digestion, the starting fresh oil contained no oxidation compounds (Flax spectrum) and the slightly oxidized contained *E,E*- and *Z,E*-conjugated dienic systems associated with hydroperoxy groups (see signals **b** and **c** in FlaxOx in Fig. (**6**)) and also a small amount of monoepoxy-octadecadienoates (see signal **e**, partially overlapped with the side band of bis-allylic protons signals **G** and **H**). After digestion, in DigFlax spectrum incipient signals **c** appeared, and in DigFlaxOx spectrum signals **b**, and especially signals **c** and **e**, increased their intensities, and new signals appeared: **a** and **d** due to conjugated dienes related to hydroxides, signal **l** due to alkanals, and signals **f**, **g** and **h**, that might be tentatively related to structures containing two or three epoxy groups.

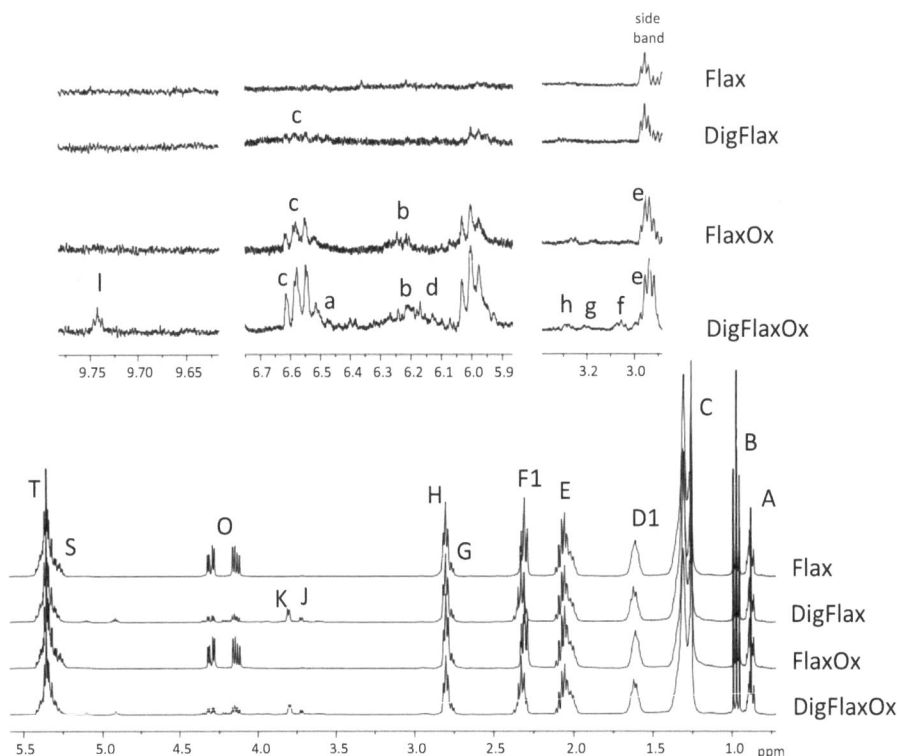

Fig. (6). ¹H NMR spectra of fresh and slightly oxidized flaxseed oils (Flax, FlaxOx) together with the lipid extracts obtained after their *in vitro* digestion (DigFlax, DigFlaxOx). Some spectral regions, where the most significant changes can be detected, are enlarged. The assignment of the signals is in agreement with Tables **1** and **2**.

For the quantitative study by ¹H NMR of digested flaxseed oil samples (DigFlax, DigFlaxOx) not only the above-described equations [eq.29-32] were used, but also the following ones to determine epoxides and aldehydes:

$$Epoxides\ (mmol/molAG+FA)=1000*(Pc*A_e/2)/NT_{AG+FA} \qquad (33)$$

$$Aldehyde\ (mmol/molAG+FA)=1000*(Pc*A_{Ald})/NT_{AG+FA} \qquad (34)$$

where A_e is the area of signal e related to monoepoxides and A_{Ald} is the area due to the aldehydic proton of the corresponding aldehyde.

Example 3: Effect of the Presence of Protein on the Oxidation of Sunflower and Flaxseed Oil During In Vitro Digestion

As a third example, Fig. (**7**) shows the ¹H NMR spectra of the lipid extracts obtained after digestion of slightly oxidized sunflower and flaxseed oils in the absence (DigSunOx, DigFlaxOx) and in the presence of proteins (DigSunOx+Prot, DigFlaxOx+Prot) [65]. As can be observed, when slightly oxidized sunflower oil was digested in the presence of ovalbumin, a clear increase of the proton signals of conjugated dienes related to hydroxides (see signals **a** and **d**) and decrease of those related to hydroperoxides (see signal **c** and lack of signal **b**) were observed (compare DigSunOx and DigSunOx+Prot spectra).

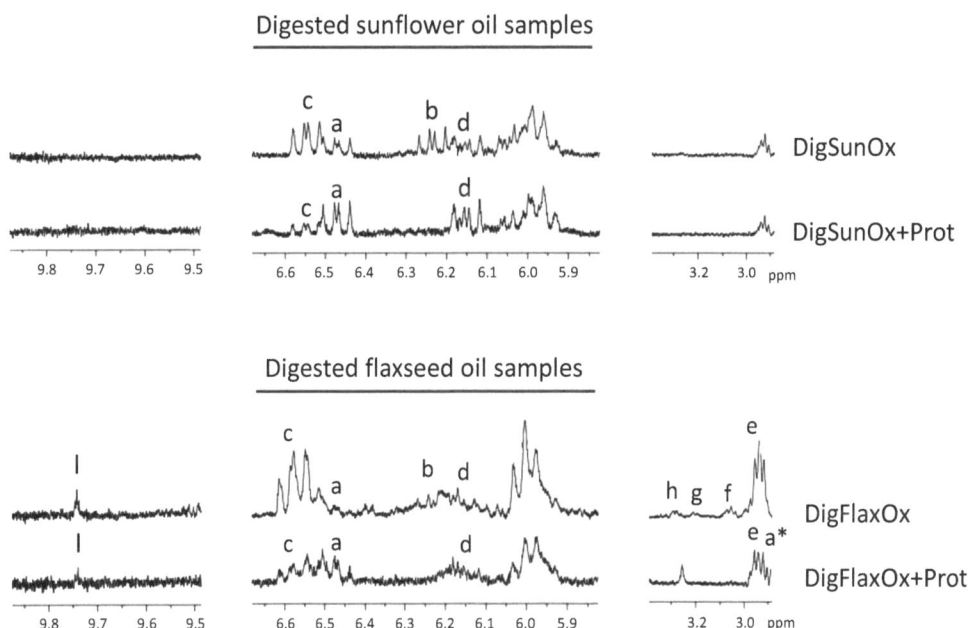

Fig. (7). ¹H NMR spectra of the lipid extracts obtained after *in vitro* digestion of slightly oxidized sunflower and flaxseed oil in the absence (DigSunOx, DigFlaxOx) and in the presence of the proteins ovalbumin (DigSunOx+Prot) and soy protein (DigFlaxOx+Prot). Some spectral regions, where the most significant changes can be detected, are enlarged. The assignment of the signals is in agreement with Tables **1** and **2**.

When slightly oxidized flaxseed oil was digested in the presence of soy protein, as in the case of sunflower oil plus ovalbumin as previously described, the advance of lipid oxidation occurred to a much lesser extent than that reached in the absence of protein (compare DigFlaxOx and DigFlaxOx+Prot spectra). Thus, the simultaneous digestion of proteins and lipids not only provoked a decrease of lipid oxidation reactions, but also provoked the occurrence of lipid reduction reactions (lipid hydroperoxides, in particular *E,E*-isomers, were almost totally reduced to more stable hydroxides). This latter result confirms the above-mentioned necessity to study as many oxidation markers as possible to assess lipid oxidation extent during digestion, because otherwise misleading conclusions can be reached. ¹H NMR spectroscopy offers this possibility, which is very difficult or even impossible to obtain with other methodologies.

Example 4: Effect of the Presence of Bioactive Compounds on the Extent of Oxidation Reactions Occurring During In Vitro Digestion

The last example on the qualitative and quantitative information provided by ¹H NMR evidences the usefulness of this technique in evaluating the effect of compounds considered as antioxidants, such as the additive 3,5-di-*tert*-butyl-4-hydroxytoluene (BHT, E-321), on the oxidation occurred during *in vitro* digestion. Fig. (**8**) shows the ¹H NMR spectra of cod liver oil non-enriched (CLO) and enriched with BHT at 800 ppm (CLO+BHT), together with the lipid extracts obtained from the corresponding *in vitro* digests (DigCLO, DigCLO+BHT) [64]. As can be observed, the commercial oil acquired had a certain oxidation degree (see in CLO spectrum signals **c** due to olefinic protons of *Z,E*-CD-OOH), but this was clearly higher after digestion (see in DigCLO the increase of the intensity of signals **c** and the appearance of signals **e** related to epoxides and **i**, **j** and **k** due to aldehydic protons of 4-hydroxy-*E*-2-alkenals, 4-hydroperoxy-*E*-2-alkenals and *Z,E*-2,4-alkadienals respectively). It is worth noting the formation of 4-hydroxy- and 4-hydroperoxy-*E*-2-alkenals as these toxic oxygenated *alpha,beta*-unsaturated aldehydes have been considered as potential causal agents of several degenerative diseases [103]. However, as can be observed in Fig. (**8**), the intensity of the above-mentioned signals due to oxidation compounds (**c**, **e**, **i**, **j** and **k**) was much lower or was absent in DigCLO+BHT spectrum in comparison with that of DigCLO Therefore, it was concluded that the addition of BHT inhibited the advance of oxidation of CLO during digestion.

Finally, summarizing all the information provided by ¹H NMR on the nature and concentrations of lipid oxidation compounds formed during *in vitro* digestion [62 - 65], it can be concluded that caution must be taken when selecting oxidation compounds as targets of analysis (markers) for assessing lipid oxidation extent: On the one hand, depending on the oxidative conditions, the kind of lipid involved, and the presence of other food components, the typical mechanism of oxidation process might not properly fit. On the other hand, erroneous conclusions concerning the oxidative status of a digested sample might be reached when determining the amount of only one or two kinds of lipid oxidation products by classical methodologies. In this context, the best option is to use innovative techniques, like ¹HNMR. Without any sample modification, these allow the simultaneous determination of a wide range of lipid oxidation products, providing as global a view of the real oxidative status of the sample as possible. It must be noted that although ¹H NMR has been employed to date to study samples from *in vitro* digestion studies (examples 1 to 4), this technique could also be successfully employed with *in vivo* samples obtained from gastric or intestinal aspirates.

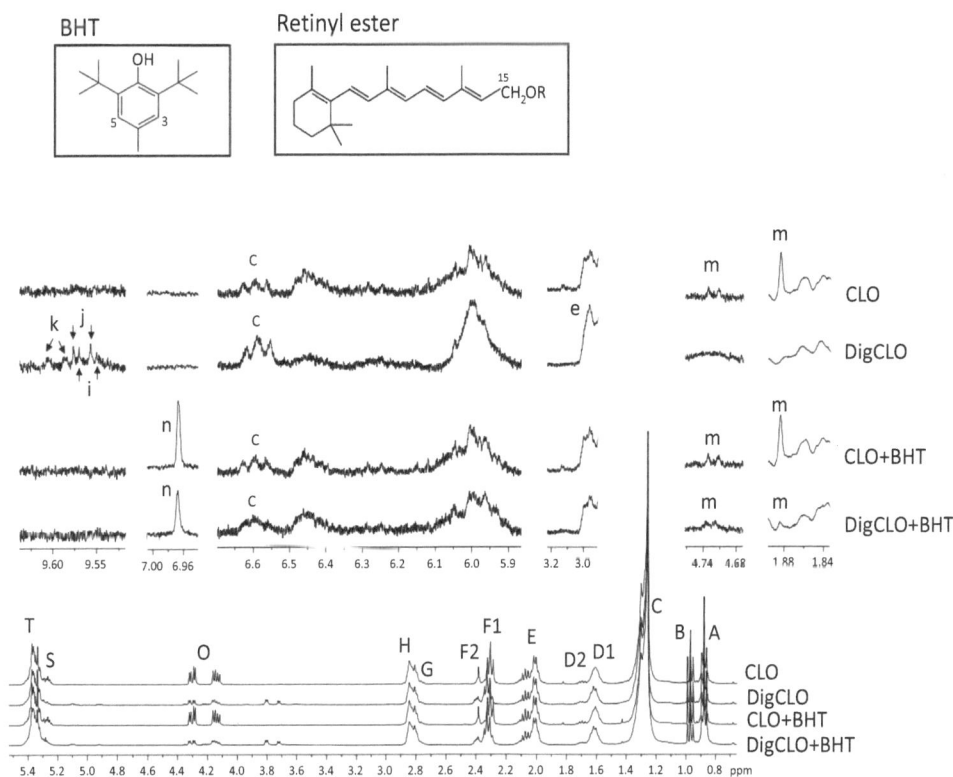

Fig. (8). ¹H NMR spectra of cod liver oil non-enriched (CLO) and enriched with the antioxidant BHT at 800 ppm (CLO+BHT), together with the lipid extracts obtained from the corresponding *in vitro* digests (DigCLO, DigCLO+BHT). Some spectral regions, where the most significant changes can be detected, are enlarged. The assignment of the signals is in agreement with Tables **1** and **2**.

3.3.3. Estimation of in vitro Bioaccesibility of Lipid Main and Minor Components and Oxidation Products

The potential application of ¹H NMR to determining the *in vitro* bioaccessibility of food lipids (L_{BA}% parameter in section 2.4.3) and of lipid oxidation products (either present in the sample before digestion or generated during this process) has been recently shown in some studies, as previously discussed. In addition to the bioaccessibility of TG (main components of food lipids), this technique also enables the study of the bioaccessibility of minor lipidic components (added or naturally present). This issue might be of particular interest nowadays, considering the increasing number of studies focusing on the fate of bioactive and/or toxic compounds during digestion.

It must be remembered that bioaccessibility is defined as the quantity of a compound that is released from its matrix in the gastrointestinal tract, becoming available for absorption. So, to estimate the bioaccessibility of any compound in a general way, the following equation based on ^{1}H NMR spectral data could be used:

$$\text{Bioaccessibility of compound X (\%)} = 100 * C_f / C_i \tag{35}$$

where C_f and C_i are the molar concentrations of the corresponding compound subject of study in the lipid extract of the *in vitro* digest (final) and in the lipidic sample before digestion (initial) respectively.

The above-mentioned concentrations C_f and C_i can be easily determined using the area of a spectral signal that does not overlap with other proton signals generated by the other components of the lipid sample (A_X) and taking into account the number of protons per molecule of the compound (n° H) that generates that signal.

$$C_X \text{ (mol/mol AG+FA)} = Pc * (A_X / n° H) / NT_{AG+FA} \tag{36}$$

where Pc is the proportionality constant existing between the area of the signals in a ^{1}H NMR spectrum and the number of protons that generate them (which could be easily determined by adding a known amount of an internal standard) and NT_{AG+FA} is the total number of moles of AG plus FA calculated by using [eq.16].

For example, these equations have been employed to estimate the bioaccessibility during *in vitro* digestion of cod liver oil of the added antioxidant BHT (see signal **n** in Fig. (**8**) at 6.97 ppm due to the two aromatic protons in C-3 and C-5 of BHT) and of naturally present vitamin A (see signal **m** in Fig. (**8**) at 4.72 ppm due to the two methylenic protons in *alpha*-position in C-15 in relation to the carbonyl group of retinyl esters) [64]. Without further enlargement it is evident that the intensity of both signals decreased after digestion (compare signal **n** in CLO+BHT spectrum with that in DigCLO+BHT, and also signal **m** in CLO spectrum with that in DigCLO). It must be noted that these equations could be applied to determine the bioaccessibility of any other lipidic component whose proton signal/s are not overlapped.

4. CONCLUSIONS AND FUTURE PERSPECTIVES

In summary, one-dimensional ¹H NMR spectroscopy has proved to be a very powerful tool in lipid digestion studies. This technique provides very valuable qualitative and quantitative information about the different types of protons present in food lipids, which can be supported on main and minor compounds, thus enabling the study of the components of a lipid sample as a whole. In a single fast run and without any sample modification, it allows the simultaneous determination of the hydrolysis and oxidation degree of any digested lipid sample by providing a great deal of qualitative and quantitative information regarding the several lipolytic species (fatty acids, glycerol. monoglycerides, diglyderides, triglyecerides), as well as a broad range of primary and secondary oxidation products present. Given the small number of studies and the limitations of the methodologies usually employed in digestion studies, ¹H NMR has contributed relevantly to shed light on the reactions affecting lipids that take place during digestion. Furthermore, it must be noted that the advantages of ¹H NMR for the evaluation of lipid hydrolysis and oxidation degree should not only be considered in the fields of food technology and nutrition, but also in many other scientific fields, such as medicine, biochemistry, biotechnology, petrochemistry and others.

As for future perspectives on the application of NMR spectroscopy in the field of lipid digestion, the increase of ¹H NMR sensitivity should be tackled specially when studying the bioaccessibility of minor lipidic compounds of interest. Indeed, these compounds, either naturally present in food or added with a nutritional purpose, might be present in digests at very low concentrations, being not possible their detection by the current ¹H NMR experiments carried out. The use of other pulse sequences which selectively suppress certain strong lipid signals could notably increase the sensitivity obtained from experiments with standard single pulse. Likewise, the application of other nucleus NMR or even advanced NMR techniques should be also encouraged in future studies dealing with food lipid digestion, because they can provide potential information that one dimensional ¹H NMR cannot. As only these latter studies have been performed to date, the possibilities offered by the above-mentioned more advanced NMR experiments remain to be explored.

CONSENT FOR PUBLICATION

Not applicable.

CONFLICT OF INTEREST

The authors declare no conflict of interest, financial or otherwise.

ACKNOWLEDGMENTS

This work has been funded by the Spanish Ministry of Economy and Competitiveness (MINECO, AGL2015-65450-R) and by the Basque Government and its Departments of Universities and Research (EJ-GV, IT-916-16) and of Economic Development and Infrastructures, Area of Agriculture, Fisheries and Food Policy (EJ-GV, PA18/04).

REFERENCES

[1]   Johnson, L.F.; Shoolery, J.N. Determination of unsaturation and average molecular weight of natural fats by nuclear magnetic resonance. *Anal. Chem.,* **1962**, *34*(9), 1136-1139.
[http://dx.doi.org/10.1021/ac60189a033]

[2]   Gunstone, F.D.; Shukla, V.K.S. NMR of lipids. In: *Annual Reports on NMR Spectroscopy: Special Edition Food Science*; Webb, GA; Belton, PS; Mccarthy, MJ, Eds.; Academic Press Inc: San Diego, **1995**; 31, pp. 219-237.
[http://dx.doi.org/10.1016/S0066-4103(08)60150-7]

[3]   Guillén, M.D.; Ruiz, A. High resolution ^{1}H nuclear magnetic resonance in the study of edible oils and fats. *Trends Food Sci. Technol.,* **2001**, *12*(9), 328-338.
[http://dx.doi.org/10.1016/S0924-2244(01)00101-7]

[4]   Mannina, L.; Sobolev, A.P.; Viel, S. Liquid state ^{1}H high field NMR in food analysis. *Prog. Nucl. Magn. Reson. Spectrosc.,* **2012**, *66*, 1-39.
[http://dx.doi.org/10.1016/j.pnmrs.2012.02.001] [PMID: 22980032]

[5]   Alexandri, E.; Ahmed, R.; Siddiqui, H.; Choudhary, M.I.; Tsiafoulis, C.G.; Gerothanassis, I.P. High Resolution NMR Spectroscopy as a Structural and Analytical Tool for Unsaturated Lipids in Solution. *Molecules,* **2017**, *22*(10), E1663.
[http://dx.doi.org/10.3390/molecules22101663] [PMID: 28981459]

[6]   Bruch, M. *NMR Spectroscopy Techniques*; CRC Press: New York, **1996**.

[7]   Hwang, H-S. *Advances in NMR Spectroscopy for Lipid Oxidation Assessment*; Springer International Publishing: Cham, Switzerland, **2017**, pp. 15-41.
[http://dx.doi.org/10.1007/978-3-319-54196-9_3]

[8]   Martínez-Yusta, A.; Goicoechea, E.; Guillén, M.D. A review of thermo-oxidative degradation of food lipids studied by ^{1}H NMR spectroscopy: Influence of degradative conditions and food lipid nature. *Compr. Rev. Food Sci. Food Saf.,* **2014**, *13*(5), 838-859.
[http://dx.doi.org/10.1111/1541-4337.12090]

[9]   Favé, G.; Coste, T.C.; Armand, M. Physicochemical properties of lipids: new strategies to manage fatty acid bioavailability. *Cell. Mol. Biol.,* **2004**, *50*(7), 815-831.
[PMID: 15672466]

[10]   McClements, D.J.; Decker, E.A.; Park, Y. Controlling lipid bioavailability through physicochemical and structural approaches. *Crit. Rev. Food Sci. Nutr.,* **2009**, *49*(1), 48-67.
[http://dx.doi.org/10.1080/10408390701764245] [PMID: 18949598]

[11]   Meynier, A.; Génot, C. Molecular and structural organization of lipids in foods: their fate during digestion and impact in nutrition. *OCL,* **2017**, *24*(2), D202.
[http://dx.doi.org/10.1051/ocl/2017006]

[12]   McClements, D.J. The biophysics of digestion: lipids. *Curr. Opin. Food Sci.,* **2018**, *21*, 1-6.
[http://dx.doi.org/10.1016/j.cofs.2018.03.009]

[13]   Carriere, F.; Barrowman, J.A.; Verger, R.; Laugier, R. Secretion and contribution to lipolysis of gastric and pancreatic lipases during a test meal in humans. *Gastroenterology,* **1993**, *105*(3), 876-888.

[http://dx.doi.org/10.1016/0016-5085(93)90908-U] [PMID: 8359655]

[14]   Borgström, B.; Erlanson, C. Pancreatic lipase and co-lipase. Interactions and effects of bile salts and other detergents. *Eur. J. Biochem.,* **1973**, *37*(1), 60-68.
[http://dx.doi.org/10.1111/j.1432-1033.1973.tb02957.x] [PMID: 4738093]

[15]   Desnuelle, P.; Savary, P. Specificity of lipases. *J. Lipid Res.,* **1963**, *4*(4), 369-384.
[PMID: 14168179]

[16]   Mattson, F.H.; Beck, L.W. The specificity of pancreatic lipase for the primary hydroxyl groups of glycerides. *J. Biol. Chem.,* **1956**, *219*(2), 735-740.
[PMID: 13319295]

[17]   Mattson, F.H.; Volpenhein, R.A. The digestion and absorption of triglycerides. *J. Biol. Chem.,* **1964**, *239*(9), 2772-2777.
[PMID: 14216426]

[18]   Miettinen, T.A.; Siurala, M. Bile salts, sterols, sterol esters, glycerides and fatty acids in micellar and oil phases of intestinal contents during fat digestion in man. *Z. Klin. Chem. Klin. Biochem.,* **1971**, *9*(1), 47-52.
[PMID: 5578066]

[19]   Reiser, R.; Bryson, M.J.; Carr, M.J.; Kuiken, K.A. The intestinal absorption of triglycerides. *J. Biol. Chem.,* **1952**, *194*(1), 131-138.
[PMID: 14927600]

[20]   Borgström, B.; Tryding, N.; Westöö, G. On the extent of hydrolysis of triglyceride ester bonds in the lumen of human small intestine during digestion. *Acta Physiol. Scand.,* **1957**, *40*(2-3), 241-247.
[http://dx.doi.org/10.1111/j.1748-1716.1957.tb01493.x] [PMID: 13469530]

[21]   Borgstroem, B. Influence of bile salt, pH, and time on the action of pancreatic lipase; physiological implications. *J. Lipid Res.,* **1964**, *5*, 522-531.
[PMID: 14221095]

[22]   Armand, M.; Borel, P.; Dubois, C.; Senft, M.; Peyrot, J.; Salducci, J.; Lafont, H.; Lairon, D. Characterization of emulsions and lipolysis of dietary lipids in the human stomach. *Am. J. Physiol.,* **1994**, *266*(3 Pt 1), G372-G381.
[PMID: 8166277]

[23]   Armand, M.; Pasquier, B.; André, M.; Borel, P.; Senft, M.; Peyrot, J.; Salducci, J.; Portugal, H.; Jaussan, V.; Lairon, D. Digestion and absorption of 2 fat emulsions with different droplet sizes in the human digestive tract. *Am. J. Clin. Nutr.,* **1999**, *70*(6), 1096-1106.
[http://dx.doi.org/10.1093/ajcn/70.6.1096] [PMID: 10584056]

[24]   Pafumi, Y.; Lairon, D.; de la Porte, P.L.; Juhel, C.; Storch, J.; Hamosh, M.; Armand, M. Mechanisms of inhibition of triacylglycerol hydrolysis by human gastric lipase. *J. Biol. Chem.,* **2002**, *277*(31), 28070-28079.
[http://dx.doi.org/10.1074/jbc.M202839200] [PMID: 11940604]

[25]   Gargouri, Y.; Pieroni, G.; Rivière, C.; Lowe, P.A.; Saunière, J.F.; Sarda, L.; Verger, R. Importance of human gastric lipase for intestinal lipolysis: an in vitro study. *Biochim. Biophys. Acta,* **1986**, *879*(3), 419-423.
[http://dx.doi.org/10.1016/0005-2760(86)90234-1] [PMID: 3778930]

[26]   Borgström, B.; Dahlqvist, A.; Lundh, G.; Sjövall, J. Studies of intestinal digestion and absorption in the human. *J. Clin. Invest.,* **1957**, *36*(10), 1521-1536.
[http://dx.doi.org/10.1172/JCI103549] [PMID: 13475490]

[27]   Reis, P.; Holmberg, K.; Watzke, H.; Leser, M.E.; Miller, R. Lipases at interfaces: a review. *Adv. Colloid Interface Sci.,* **2009**, *147-148*, 237-250.
[http://dx.doi.org/10.1016/j.cis.2008.06.001] [PMID: 18691682]

[28]   Golding, M.; Wooster, T.J. The influence of emulsion structure and stability on lipid digestion. *Curr.*

Opin. Colloid Interface Sci., **2010**, *15*(1-2), 90-101.
[http://dx.doi.org/10.1016/j.cocis.2009.11.006]

[29]   Mu, H.; Høy, C.E. The digestion of dietary triacylglycerols. *Prog. Lipid Res.,* **2004**, *43*(2), 105-133.
[http://dx.doi.org/10.1016/S0163-7827(03)00050-X] [PMID: 14654090]

[30]   Capolino, P.; Guérin, C.; Paume, J.; Giallo, J.; Ballester, J.M.; Cavalier, J.F.; Carrière, F. *In vitro*
gastrointestinal lipolysis: Replacement of human digestive lipases by a combination of rabbit gastric
and porcine pancreatic extracts. *Food Dig.,* **2011**, *2*(1-3), 43-51.
[http://dx.doi.org/10.1007/s13228-011-0014-5]

[31]   Sek, L.; Porter, C.J.; Charman, W.N. Characterisation and quantification of medium chain and long
chain triglycerides and their *in vitro* digestion products, by HPTLC coupled with *in situ* densitometric
analysis. *J. Pharm. Biomed. Anal.,* **2001**, *25*(3-4), 651-661.
[http://dx.doi.org/10.1016/S0731-7085(00)00528-8] [PMID: 11377046]

[32]   Vinarov, Z.; Petkova, Y.; Tcholakova, S.; Denkov, N.; Stoyanov, S.; Pelan, E.; Lips, A. Effects of
emulsifier charge and concentration on pancreatic lipolysis. 1. In the absence of bile salts. *Langmuir,*
2012, *28*(21), 8127-8139.
[http://dx.doi.org/10.1021/la300366m] [PMID: 22554275]

[33]   Vinarov, Z.; Tcholakova, S.; Damyanova, B.; Atanasov, Y.; Denkov, N.D.; Stoyanov, S.D.; Pelan, E.;
Lips, A. Effects of emulsifier charge and concentration on pancreatic lipolysis: 2. Interplay of
emulsifiers and biles. *Langmuir,* **2012**, *28*(33), 12140-12150.
[http://dx.doi.org/10.1021/la301820w] [PMID: 22680619]

[34]   Martin, D.; Nieto-Fuentes, J.A.; Señoráns, F.J.; Reglero, G.; Soler-Rivas, C. Intestinal digestion of fish
oils and ω-3 concentrates under *in vitro* conditions. *Eur. J. Lipid Sci. Technol.,* **2010**, *112*(12), 1315-
1322.
[http://dx.doi.org/10.1002/ejlt.201000329]

[35]   Kenmogne-Domguia, H.B.; Meynier, A.; Viau, M.; Llamas, G.; Genot, C. Gastric conditions control
both the evolution of the organization of protein-stabilized emulsions and the kinetic of lipolysis
during *in vitro* digestion. *Food Funct.,* **2012**, *3*(12), 1302-1309.
[http://dx.doi.org/10.1039/c2fo30031a] [PMID: 22918290]

[36]   Helbig, A.; Silletti, E.; Timmerman, E.; Hamer, R.J.; Gruppen, H. *In vitro* study of intestinal lipolysis
using pH-stat and gas chromatography. *Food Hydrocoll.,* **2012**, *28*(1), 10-19.
[http://dx.doi.org/10.1016/j.foodhyd.2011.11.007]

[37]   Shen, Z.; Apriani, C.; Weerakkody, R.; Sanguansri, L.; Augustin, M.A. Food matrix effects on in vitro
digestion of microencapsulated tuna oil powder. *J. Agric. Food Chem.,* **2011**, *59*(15), 8442-8449.
[http://dx.doi.org/10.1021/jf201494b] [PMID: 21721584]

[38]   Ye, A.; Cui, J.; Zhu, X.; Singh, H. Effect of calcium on the kinetics of free fatty acid release during in
vitro lipid digestion in model emulsions. *Food Chem.,* **2013**, *139*(1-4), 681-688.
[http://dx.doi.org/10.1016/j.foodchem.2013.02.014] [PMID: 23561161]

[39]   Zhu, X.; Ye, A.; Verrier, T.; Singh, H. Free fatty acid profiles of emulsified lipids during *in vitro*
digestion with pancreatic lipase. *Food Chem.,* **2013**, *139*(1-4), 398-404.
[http://dx.doi.org/10.1016/j.foodchem.2012.12.060] [PMID: 23561123]

[40]   Fatouros, D.G.; Bergenstahl, B.; Mullertz, A. Morphological observations on a lipid-based drug
delivery system during in vitro digestion. *Eur. J. Pharm. Sci.,* **2007**, *31*(2), 85-94.
[http://dx.doi.org/10.1016/j.ejps.2007.02.009] [PMID: 17418543]

[41]   Brogård, M.; Troedsson, E.; Thuresson, K.; Ljusberg-Wahren, H. A new standardized lipolysis
approach for characterization of emulsions and dispersions. *J. Colloid Interface Sci.,* **2007**, *308*(2),
500-507.
[http://dx.doi.org/10.1016/j.jcis.2006.12.015] [PMID: 17289070]

[42]   Li, Y.; McClements, D.J. New mathematical model for interpreting pH-stat digestion profiles: impact

of lipid droplet characteristics on in vitro digestibility. *J. Agric. Food Chem.,* **2010**, *58*(13), 8085-8092.
[http://dx.doi.org/10.1021/jf101325m] [PMID: 20557040]

[43] Thomas, N.; Holm, R.; Rades, T.; Müllertz, A. Characterising lipid lipolysis and its implication in lipid-based formulation development. *AAPS J.,* **2012**, *14*(4), 860-871.
[http://dx.doi.org/10.1208/s12248-012-9398-6] [PMID: 22956477]

[44] Marze, S.; Meynier, A.; Anton, M. In vitro digestion of fish oils rich in n-3 polyunsaturated fatty acids studied in emulsion and at the oil-water interface. *Food Funct.,* **2013**, *4*(2), 231-239.
[http://dx.doi.org/10.1039/C2FO30165B] [PMID: 23086175]

[45] Sek, L.; Porter, C.J.; Kaukonen, A.M.; Charman, W.N. Evaluation of the in-vitro digestion profiles of long and medium chain glycerides and the phase behaviour of their lipolytic products. *J. Pharm. Pharmacol.,* **2002**, *54*(1), 29-41.
[http://dx.doi.org/10.1211/0022357021771896] [PMID: 11833493]

[46] Kanicky, J.R.; Shah, D.O. Effect of premicellar aggregation on the pKa of fatty acid soap solution. *Langmuir,* **2003**, *19*, 2034-2038.
[http://dx.doi.org/10.1021/la020672y]

[47] Nieva-Echevarría, B.; Goicoechea, E.; Manzanos, M.J.; Guillén, M.D. A method based on ¹H NMR spectral data useful to evaluate the hydrolysis level in complex lipid mixtures. *Food Res. Int.,* **2014**, *66*, 379-387.
[http://dx.doi.org/10.1016/j.foodres.2014.09.031]

[48] Nieva-Echevarría, B.; Goicoechea, E.; Manzanos, M.J.; Guillén, M.D. Usefulness of (¹)H NMR in assessing the extent of lipid digestion. *Food Chem.,* **2015**, *179*, 182-190.
[http://dx.doi.org/10.1016/j.foodchem.2015.01.104] [PMID: 25722153]

[49] Guillén, M.D.; Ruiz, A. Rapid simultaneous determination by proton NMR of unsaturation and composition of acyl groups in vegetable oils. *Eur. J. Lipid Sci. Technol.,* **2003**, *105*, 688-696.
[http://dx.doi.org/10.1002/ejlt.200300866]

[50] Guillén, M.D.; Ruiz, A. Study of the oxidative stability of salted and unsalted salmon fillets by ¹H nuclear magnetic resonance. *Food Chem.,* **2004**, *86*(2), 297-304.
[http://dx.doi.org/10.1016/j.foodchem.2003.09.028]

[51] Guillén, M.D.; Goicoechea, E. Detection of primary and secondary oxidation products by Fourier transform infrared spectroscopy (FTIR) and ¹H nuclear magnetic resonance (NMR) in sunflower oil during storage. *J. Agric. Food Chem.,* **2007**, *55*(26), 10729-10736.
[http://dx.doi.org/10.1021/jf071712c] [PMID: 18038977]

[52] Guillén, M.D.; Uriarte, P.S. Contribution to further understanding of the evolution of sunflower oil submitted to frying temperature in a domestic fryer: study by ¹H nuclear magnetic resonance. *J. Agric. Food Chem.,* **2009**, *57*(17), 7790-7799.
[http://dx.doi.org/10.1021/jf900510k] [PMID: 19663427]

[53] Guillén, M.D.; Uriarte, P.S. Study by ¹H NMR spectroscopy of the evolution of extra virgin olive oil composition submitted to frying temperature in an industrial fryer for a prolonged period of time. *Food Chem.,* **2012**, *134*(1), 162-172.
[http://dx.doi.org/10.1016/j.foodchem.2012.02.083]

[54] Martínez-Yusta, A.; Guillén, M.D. Deep-frying food in extra virgin olive oil: a study by (¹)H nuclear magnetic resonance of the influence of food nature on the evolving composition of the frying medium. *Food Chem.,* **2014**, *150*, 429-437.
[http://dx.doi.org/10.1016/j.foodchem.2013.11.015] [PMID: 24360472]

[55] Sacchi, R.; Paolillo, L.; Giudicianni, I.; Addeo, F. Rapid ¹H-NMR determination of 1,2 and 1,3 diglycerides in virgin olive oils. *Ital. J. Food Sci.,* **1991**, (4), 253-262.

[56] Compton, D.L.; Vermillion, K.E.; Laszlo, J.A. Acyl migration kinetics of 2-monoacylglycerols from soybean oil *via*¹H NMR. *J. Am. Oil Chem. Soc.,* **2007**, *84*(4), 343-348.

[http://dx.doi.org/10.1007/s11746-007-1049-1]

[57]   Jin, F.; Kawasaki, K.; Kishida, H.; Tohji, K.; Moriya, T.; Enomoto, H. NMR spectroscopic study on methanolysis reaction of vegetable oil. *Fuel,* **2007**, *86*(7-8), 1201-1207.
[http://dx.doi.org/10.1016/j.fuel.2006.10.013]

[58]   Satyarthi, J.K.; Srinivas, D.; Ratnasamy, P. Estimation of free fatty acid content in oils, fats, and biodiesel by ^{1}H NMR spectroscopy. *Energy Fuels,* **2009**, *23*(4), 2273-2277.
[http://dx.doi.org/10.1021/ef801011v]

[59]   Kumar, R.; Bansal, V.; Tiwari, A.K.; Sharma, M.; Puri, S.K.; Patel, M.B.; Sarpal, A.S. Estimation of glycerides and free fatty acid in oils extracted from various seeds from the Indian region by NMR spectroscopy. *J. Am. Oil Chem. Soc.,* **2011**, *88*(11), 1675-1685.
[http://dx.doi.org/10.1007/s11746-011-1846-4]

[60]   Skiera, C.; Steliopoulos, P.; Kuballa, T.; Holzgrabe, U.; Diehl, B. Determination of free fatty acids in edible oils by ^{1}H NMR spectroscopy. *Lipid Technol.,* **2012**, *24*(12), 279-281.
[http://dx.doi.org/10.1002/lite.201200241]

[61]   Sopelana, P.; Arizabaleta, I.; Ibargoitia, M.L.; Guillén, M.D. Characterisation of the lipidic components of margarines by ^{1}H Nuclear Magnetic Resonance. *Food Chem.,* **2013**, *141*(4), 3357-3364.
[http://dx.doi.org/10.1016/j.foodchem.2013.06.026] [PMID: 23993493]

[62]   Nieva-Echevarría, B.; Goicoechea, E.; Manzanos, M.J.; Guillén, M.D. ^{1}H NMR and SPME-GC/MS study of hydrolysis, oxidation and other reactions occurring during in vitro digestion of non-oxidized and oxidized sunflower oil. Formation of hydroxy-octadecadienoates. *Food Res. Int.,* **2017**, *91*, 171-182.
[http://dx.doi.org/10.1016/j.foodres.2016.11.027] [PMID: 28290321]

[63]   Nieva-Echevarría, B.; Goicoechea, E.; Guillén, M.D. Behaviour of non-oxidized and oxidized flaxseed oils, as models of omega-3 rich lipids, during *in vitro* digestion. Occurrence of epoxidation reactions. *Food Res. Int.,* **2017**, *97*, 104-115.
[http://dx.doi.org/10.1016/j.foodres.2017.03.047] [PMID: 28578030]

[64]   Nieva-Echevarría, B.; Goicoechea, E.; Guillén, M.D. Polyunsaturated lipids and vitamin A oxidation during cod liver oil *in vitro* gastrointestinal digestion. Antioxidant effect of added BHT. *Food Chem.,* **2017**, *232*, 733-743.
[http://dx.doi.org/10.1016/j.foodchem.2017.04.057] [PMID: 28490135]

[65]   Nieva-Echevarría, B.; Goicoechea, E.; Guillén, M.D. Effect of the presence of protein on lipolysis and lipid oxidation occurring during *in vitro* digestion of highly unsaturated oils. *Food Chem.,* **2017**, *235*, 21-33.
[http://dx.doi.org/10.1016/j.foodchem.2017.05.028] [PMID: 28554628]

[66]   Nieva-Echevarría, B.; Goicoechea, E.; Manzanos, M.J.; Guillén, M.D. Fish *in vitro* digestion: Influence of fish salting on the extent of lipolysis, oxidation and other reactions. *J. Agric. Food Chem.,* **2017**, *65*(4), 879-891.
[http://dx.doi.org/10.1021/acs.jafc.6b04334] [PMID: 28052192]

[67]   Nieva-Echevarría, B.; Goicoechea, E.; Guillén, M.D. Effect of liquid smoking on lipid hydrolysis and oxidation reactions during *in vitro* gastrointestinal digestion of European sea bass. *Food Res. Int.,* **2017**, *97*, 51-61.
[http://dx.doi.org/10.1016/j.foodres.2017.03.032] [PMID: 28578064]

[68]   Rodriguez, J.A.; Mendoza, L.D.; Pezzotti, F.; Vanthuyne, N.; Leclaire, J.; Verger, R.; Buono, G.; Carriere, F.; Fotiadu, F. Novel chromatographic resolution of chiral diacylglycerols and analysis of the stereoselective hydrolysis of triacylglycerols by lipases. *Anal. Biochem.,* **2008**, *375*(2), 196-208.
[http://dx.doi.org/10.1016/j.ab.2007.11.036] [PMID: 18162167]

[69]   Li, Y.; Hu, M.; McClements, D.J. Factors affecting lipase digestibility of emulsified lipids using an *in vitro* digestion model: Proposal for a standardised pH-stat method. *Food Chem.,* **2011**, *126*(2), 498-

505.
[http://dx.doi.org/10.1016/j.foodchem.2010.11.027]

[70]   Lamothe, S.; Corbeil, M.M.; Turgeon, S.L.; Britten, M. Influence of cheese matrix on lipid digestion in a simulated gastro-intestinal environment. *Food Funct.,* **2012**, *3*(7), 724-731.
[http://dx.doi.org/10.1039/c2fo10256k] [PMID: 22476332]

[71]   Halliwell, B.; Zhao, K.; Whiteman, M. The gastrointestinal tract: a major site of antioxidant action? *Free Radic. Res.,* **2000**, *33*(6), 819-830.
[http://dx.doi.org/10.1080/10715760000301341] [PMID: 11237104]

[72]   Holman, R.T.; Elmer, O.C. The rates of oxidation of unsaturated fatty acids and esters. *J. Am. Oil Chem. Soc.,* **1947**, *24*(4), 127-129.
[http://dx.doi.org/10.1007/BF02643258]

[73]   Kanner, J.; Lapidot, T. The stomach as a bioreactor: dietary lipid peroxidation in the gastric fluid and the effects of plant-derived antioxidants. *Free Radic. Biol. Med.,* **2001**, *31*(11), 1388-1395.
[http://dx.doi.org/10.1016/S0891-5849(01)00718-3] [PMID: 11728810]

[74]   Gorelik, S.; Lapidot, T.; Shaham, I.; Granit, R.; Ligumsky, M.; Kohen, R.; Kanner, J. Lipid peroxidation and coupled vitamin oxidation in simulated and human gastric fluid inhibited by dietary polyphenols: health implications. *J. Agric. Food Chem.,* **2005**, *53*(9), 3397-3402.
[http://dx.doi.org/10.1021/jf040401o] [PMID: 15853378]

[75]   Goicoechea, E.; Van Twillert, K.; Duits, M.; Brandon, E.D.F.A.; Kootstra, P.R.; Blokland, M.H.; Guillén, M.D. Use of an *in vitro* digestion model to study the bioaccessibility of 4-hydroxy-2-nonenal and related aldehydes present in oxidized oils rich in omega-6 acyl groups. *J. Agric. Food Chem.,* **2008**, *56*(18), 8475-8483.
[http://dx.doi.org/10.1021/jf801212k] [PMID: 18729379]

[76]   Kenmogne-Domguia, H.B.; Meynier, A.; Boulanger, C.; Genot, C. Lipid oxidation in food emulsions under gastrointestinal-simulated conditions: the key role of endogenous tocopherols and initiator. *Food Dig.,* **2012**, *3*(1-3), 46-52.
[http://dx.doi.org/10.1007/s13228-012-0026-9]

[77]   Kenmogne-Domguia, H.B.; Moisan, S.; Viau, M.; Genot, C.; Meynier, A. The initial characteristics of marine oil emulsions and the composition of the media inflect lipid oxidation during *in vitro* gastrointestinal digestion. *Food Chem.,* **2014**, *152*, 146-154.
[http://dx.doi.org/10.1016/j.foodchem.2013.11.096] [PMID: 24444919]

[78]   Van Hecke, T.; Vossen, E.; Hemeryck, L.Y.; Vanden Bussche, J.; Vanhaecke, L.; De Smet, S. Increased oxidative and nitrosative reactions during digestion could contribute to the association between well-done red meat consumption and colorectal cancer. *Food Chem.,* **2015**, *187*, 29-36.
[http://dx.doi.org/10.1016/j.foodchem.2015.04.029] [PMID: 25976994]

[79]   Frankel, E.N. *Lipid Oxidation,* 2nd ed; Woodhead Publishing Limited: Cambridge, **2005**.
[http://dx.doi.org/10.1533/9780857097927]

[80]   Schaich, K.M. Challenges in elucidating lipid oxidation mechanisms: when, where, and how do products arise? In: *Lipid oxidation: Challenges in food systems*; Logan, A.; Nienaber, U.; Pan, X., Eds.; AOCS Press: Oxford, UK, **2013**; pp. 1-52.
[http://dx.doi.org/10.1016/B978-0-9830791-6-3.50004-7]

[81]   Schaich, K.M. Analysis of lipid and protein oxidation in fats, oils, and foods. In: *Oxidative stability and shelf life of foods containing oils and fats*; Hu, M.; Jacobsen, C., Eds.; Academic Press and AOCS Press: Oxford, UK, **2016**; pp. 1-131.
[http://dx.doi.org/10.1016/B978-1-63067-056-6.00001-X]

[82]   Martin-Rubio, A.S.; Sopelana, P.; Ibargoitia, M.L.; Guillén, M.D. Prooxidant effect of α-tocopherol on soybean oil. Global monitoring of its oxidation process under accelerated storage conditions by ¹H nuclear magnetic resonance. *Food Chem.,* **2018**, *245*, 312-323.
[http://dx.doi.org/10.1016/j.foodchem.2017.10.098] [PMID: 29287377]

[83] Kerem, Z.; Chetrit, D.; Shoseyov, O.; Regev-Shoshani, G. Protection of lipids from oxidation by epicatechin, *trans*-resveratrol, and gallic and caffeic acids in intestinal model systems. *J. Agric. Food Chem.,* **2006**, *54*(26), 10288-10293.
[http://dx.doi.org/10.1021/jf0621828] [PMID: 17177572]

[84] Kuffa, M.; Priesbe, T.J.; Krueger, C.G.; Reed, J.D.; Richards, M.P. Ability of dietary antioxidants to affect lipid oxidation of cooked turkey meat in a simulated stomach and blood lipids after a meal. *J. Funct. Foods,* **2009**, *1*(2), 208-216.
[http://dx.doi.org/10.1016/j.jff.2009.01.010]

[85] Lorrain, B.; Dangles, O.; Loonis, M.; Armand, M.; Dufour, C. Dietary iron-initiated lipid oxidation and its inhibition by polyphenols in gastric conditions. *J. Agric. Food Chem.,* **2012**, *60*(36), 9074-9081.
[http://dx.doi.org/10.1021/jf302348s] [PMID: 22860567]

[86] Steppeler, C.; Haugen, J.E.; Rødbotten, R.; Kirkhus, B. Formation of malondialdehyde, 4-hydroxynonenal, and 4-hydroxyhexenal during *in vitro* digestion of cooked beef, pork, chicken, and salmon. *J. Agric. Food Chem.,* **2016**, *64*(2), 487-496.
[http://dx.doi.org/10.1021/acs.jafc.5b04201] [PMID: 26654171]

[87] Tullberg, C.; Larsson, K.; Carlsson, N.G.; Comi, I.; Scheers, N.; Vegarud, G.; Undeland, I. Formation of reactive aldehydes (MDA, HHE, HNE) during the digestion of cod liver oil: comparison of human and porcine *in vitro* digestion models. *Food Funct.,* **2016**, *7*(3), 1401-1412.
[http://dx.doi.org/10.1039/C5FO01332A] [PMID: 26838473]

[88] Goicoechea, E.; Brandon, E.F.A.; Blokland, M.H.; Guillén, M.D. Fate in digestion *in vitro* of several food components, including some toxic compounds coming from omega-3 and omega-6 lipids. *Food Chem. Toxicol.,* **2011**, *49*(1), 115-124.
[http://dx.doi.org/10.1016/j.fct.2010.10.005] [PMID: 20937346]

[89] Claxson, A.W.; Hawkes, G.E.; Richardson, D.P.; Naughton, D.P.; Haywood, R.M.; Chander, C.L.; Atherton, M.; Lynch, E.J.; Grootveld, M.C. Generation of lipid peroxidation products in culinary oils and fats during episodes of thermal stressing: a high field ^{1}H NMR study. *FEBS Lett.,* **1994**, *355*(1), 81-90.
[http://dx.doi.org/10.1016/0014-5793(94)01147-8] [PMID: 7957968]

[90] Haywood, R.M.; Claxson, A.W.D.; Hawkes, G.E.; Richardson, D.P.; Naughton, D.P.; Coumbarides, G.; Hawkes, J.; Lynch, E.J.; Grootveld, M.C. Detection of aldehydes and their conjugated hydroperoxydiene precursors in thermally-stressed culinary oils and fats: investigations using high resolution proton NMR spectroscopy. *Free Radic. Res.,* **1995**, *22*(5), 441-482.
[http://dx.doi.org/10.3109/10715769509147552] [PMID: 7633572]

[91] Guillén, M.D.; Goicoechea, E. Oxidation of corn oil at room temperature: Primary and secondary oxidation products and determination of their concentration in the oil liquid matrix from ^{1}H nuclear magnetic resonance data. *Food Chem.,* **2009**, *116*(1), 183-192.
[http://dx.doi.org/10.1016/j.foodchem.2009.02.029]

[92] Calvo, P.; Lozano, M.; Espinosa-Mansilla, A.; Gónzalez-Gómez, D. In vitro evaluation of the availability of ω-3 and ω-6 fatty acids and tocopherols from microencapsulated walnut oil. *Food Res. Int.,* **2012**, *48*, 316-321.
[http://dx.doi.org/10.1016/j.foodres.2012.05.007]

[93] Aursand, M.; Rainuzzo, J.R.; Grasladen, H. Quantitative high-resolution ^{13}C and ^{1}H nuclear magnetic resonance of omega-3 fatty acids from white muscle of atlantic salmon (*Salmo salar*). *J. Am. Oil Chem. Soc.,* **1993**, *70*, 971-981.
[http://dx.doi.org/10.1007/BF02543023]

[94] Aerts, A.A.J.; Jacobs, P.A. Epoxide yield determination of oils and fatty acid methyl esters using 1H NMR. *J. Am. Oil Chem. Soc.,* **2004**, *81*(9), 841-846.
[http://dx.doi.org/10.1007/s11746-004-0989-1]

[95]   Goicoechea, E.; Guillén, M.D. Analysis of hydroperoxides, aldehydes and epoxides by 1H nuclear magnetic resonance in sunflower oil oxidized at 70 and 100 degrees C. *J. Agric. Food Chem.,* **2010**, *58*(10), 6234-6245.
       [http://dx.doi.org/10.1021/jf1005337] [PMID: 20438132]

[96]   Gardner, H.W.; Weisleder, D. Hydroperoxides from oxidation of linoleic and linolenic acids by soybean lipoxygenase: Proof of the *trans*-11 double bond. *Lipids,* **1972**, *7*(3), 191-193.
       [http://dx.doi.org/10.1007/BF02533062]

[97]   Murakami, N.; Shirahashi, H.; Nagatsu, A.; Sakakibara, J. Two unsaturated 9R-hydroxy fatty acids from the cyanobacterium *Anabaena flos-aquae f. flos-aquae. Lipids,* **1992**, *27*(10), 776-778.
       [http://dx.doi.org/10.1007/BF02535848]

[98]   Kuklev, D.V.; Christie, W.W.; Durand, T.; Rossi, J.C.; Vidal, J.P.; Kasyanov, S.P.; Akulin, V.A.; Bezuglov, V.V. Synthesis of keto- and hydroxydienoic compounds from linoleic acid. *Chem. Phys. Lipids,* **1997**, *85*(2), 125-134.
       [http://dx.doi.org/10.1016/S0009-3084(96)02650-3] [PMID: 9275308]

[99]   Cui, P.H.; Duke, R.K.; Duke, C.C. Monoepoxy octadecadienoates and monoepoxy octadecatrienoates 1: NMR spectral characterization. *Chem. Phys. Lipids,* **2008**, *152*(2), 122-130.
       [http://dx.doi.org/10.1016/j.chemphyslip.2008.02.003] [PMID: 18339314]

[100] Amato, M.E.; Ballistreri, F.P.; Pappalardo, A.; Tomaselli, G.A.; Toscano, R.M.; Sfrazzetto, G.T. Selective oxidation reactions of natural compounds with hydrogen peroxide mediated by methyltrioxorhenium. *Molecules,* **2013**, *18*(11), 13754-13768.
       [http://dx.doi.org/10.3390/molecules181113754] [PMID: 24213654]

[101] Kikuchi, M.; Yaoita, Y.; Kikuchi, M. Monohydroxy-Substituted Polyunsaturated Fatty Acids from *Swertia japonica. Helv. Chim. Acta,* **2008**, *91*(10), 1857-1862.
       [http://dx.doi.org/10.1002/hlca.200890198]

[102] Tassignon, P.; de Waard, P.; de Rijk, T.; Tournois, H.; de Wit, D.; de Buyck, L. An efficient countercurrent distribution method for the large-scale isolation of dimorphecolic acid methyl ester. *Chem. Phys. Lipids,* **1994**, *71*(2), 187-196.
       [http://dx.doi.org/10.1016/0009-3084(94)90070-1] [PMID: 8033290]

[103] Guillén, M.D.; Goicoechea, E. Toxic oxygenated *alpha,beta*-unsaturated aldehydes and their study in foods: a review. *Crit. Rev. Food Sci. Nutr.,* **2008**, *48*(2), 119-136.
       [http://dx.doi.org/10.1080/10408390601177613] [PMID: 18274968]

CHAPTER 4

Nuclear Magnetic Resonance as an Attractive Resource for Monitoring Surveillance Candidates of Acute and Chronic Lung Disorders

Simona Viglio¹, Cristina Airoldi², Carlotta Ciaramelli² and Paolo Iadarola³,*

¹ Department of Molecular Medicine, University of Pavia, Via Taramelli 3, Pavia, Italy

² Department of Biotechnologies and Biosciences, University of Milano-Bicocca, Piazza della Scienza 2, Milano, Italy

³ Department of Biology and Biotechnologies "L.Spallanzani", University of Pavia, Via A.Ferrata 9, Pavia, Italy

Abstract: Metabolomics is the comprehensive study of *metabolites, i.e.* substrates and end-products of cell *metabolism*. These are low-molecular weight molecules which include amino, nucleic and organic acids, peptides, carbohydrates, vitamins, polyphenols, alkaloids and inorganic species. Being metabolite concentration influenced by both genetic and environmental factors, their amount directly reflects the underlying biochemical activity and state of cells, tissues or organisms. Profiling the metabolome could thus represent the molecular *phenotype* better than other approaches such as genomics and proteomics.

Among the available procedures (Gas Chromatography-/Liquid Chromatography-Mass Spectrometry), high-resolution nuclear magnetic resonance spectroscopy (HR-NMR) is currently one of the leading analytical tools for metabolomic research due to its peculiarities. The distinctive advantage of NMR over other methods is the possibility to perform an inherent quantitative and untargeted analysis, also with respect to the chemical nature of metabolites. In addition, NMR shows a good reproducibility, a rapid acquisition time of spectra, and it is not destructive with regard to the sample for which little or no preparation is required. Taken together, these features have promoted NMR-assisted metabolomics to the rank of a valuable method for an efficient investigation of a variety of lung diseases.

Aim of this chapter is to provide an overview of the applications of metabolomics to the study of acute and chronic lung disorders. Why focus on pulmonary disorders? First, by involving tens of million people, lung diseases are some of the most common medical conditions in the world. Second, the depth of analysis ultimately reached by current metabolomic procedures has provided a new and larger context for future studies on the biology of these conditions. This has allowed for the generation of

***** **Corresponding author Paolo Iadarola:** Department of Biology and Biotechnologies "L. Spallanzani", University of Pavia, *Via* A. Ferrata 9, 27100- Pavia, Italy; Tel: +39 0382 987264, Fax: +39 0382 423108, Email: piadarol@unipv.it

Atta-ur-Rahman & M. Iqbal Choudhary (Eds.)

metabolite profiles that could be useful for exploring pathological mechanisms and/or discovering new potential therapeutic targets for a variety of pulmonary disorders.

Keywords: Nuclear Magnetic Resonance, Pulmonary Disorders, Metabolic Profiles, Metabolic Pathways, Statistical Analyses, Profile Comparison, Up-regulation, Down-regulation, Amino Acids, Nucleosides, Nucleotides, Ketoacids, Krebs Cycle Intermediates.

INTRODUCTION TO METABOLOMICS

Unravelling the complexity of a disease with the aim to identify predictive markers may be prevented by ignorance of the specific molecular mechanisms that cause the onset of the disorder and govern its course. A poor understanding of disease processes at the metabolic level is indeed the main obstacle for obtaining their diagnosis/prognosis and for the development of a possible therapy. For this reason, researchers have recently focused their attention on metabolites, *i.e. ionic inorganic species, carbohydrates, volatile alcohols and ketones, organic and amino acids, lipids, nucleosides/nucleotides* and many more. First of all, these small molecules are the end products of transcriptional and translational processes and, as such, they can be seen as the output of the biological system, *i.e.* a molecular phenotype. Second, as perturbations in the levels and fluxes of specific endogenous intermediates of a sequential series of reactions are often more pronounced than variations in enzymatic kinetics, they may be more informative than these latter. Thus, on the assumption that metabolites contain information that is closely related to the current status of an individual, it may be argued that any alteration of their overall profile is strictly connected with changes in biological condition of individuals. In light of the above, it seems obvious that understanding the metabolic distinctive variations that accompany a given disorder would hopefully provide crucial insights into disease pathophysiology [1]. While the question of whether alterations of metabolites cause or are the consequence of a disease remains unanswered, comparing metabolic profiles of sick to healthy individuals may be a helpful approach for the identification of disease biomarkers that could be good indicators of disease onset/progression.

Hunting for the contribution of metabolite changes to a disease progression would require the use of sophisticated techniques able to provide the qualitative/ quantitative detection of a large number of metabolites within a variety of biological matrices. However, the metabolic profile of a biological sample is often affected by a variety of factors which include diet, drugs, age, lifestyle, ethnicity and others which could cause the consequent misinterpretation of data. Because these drawbacks cannot be underestimated, a strict control or deconvolution of all these factors is essential to obtain reliable information on a given disease [2, 3].

Profiling molecular signatures of disease processes and disease-relevant fluctuations of mctabolites is currently achieved by means of *metabolomics,* an efficient approach for the detection in biological matrices of up to many hundreds of metabolites. Due to the huge diversity of chemical characteristics and abundance of bioanalytes, capturing the entire metabolome released in a biological sample is not trivial. Because the identification of metabolites that discriminate different cohorts of individuals is very challenging, several procedures have been developed to achieve this goal. Despite a multiplatform approach would appear more appropriate than a single technique [4, 5], a great deal of literature in this field demonstrates that also a single analytical method is able to provide a good snapshot of the system under investigation. Liquid- or gas-chromatography mass spectrometry (LC- or GC-MS) and nuclear magnetic resonance (NMR) spectroscopy are the most successful approaches currently applied [6 - 13]. By providing complementary information, these methods are often used synergistically either to observe the metabolic state of an organism and to yield valuable information concerning the functional alterations of biochemical pathways. All of these methods are characterized by strengths and weaknesses. For example, while MS offers high sensitivity and high throughput, it also shows a good number of weaknesses. These include: i) the need to be combined with a preliminary separation step when applied to complex biological matrices, ii) difficulties with quantification of analytes, iii) problems involving the sample preparation and its molecular environment that may affect signal intensity [14]. By contrast, NMR spectroscopy shows peculiarities that make it more attractive than other techniques. First of all, it requires a minimal sample pre-treatment and, very often, no extra-steps for sample preparation/derivatization are necessary. This feature gives NMR analysis an excellent practical handling in the lab. Second, NMR is rapid, non-destructive, highly reproducible and rigorously quantitative over a wide dynamic range. Compared to MS it also offers advantages for the identification of compounds with identical masses and for analytes that are difficult to ionize or require derivatization. Lastly, given the enormous technological improvements that continue to merge, the weak point related to its non-excellent sensitivity (indeed much lower than that of MS) does not seem to represent a key-limitation of the technique. Taken together, all mentioned advantages make NMR well suited for the comprehensive and simultaneous analysis of a wide variety of compounds, *i.e.* for metabolomic analysis.

To avoid misunderstandings, the authors would like to clarify the meaning of *metabonomics*, a term that may be encountered in a few reports in this field. It differs more in practice than in definition from the other term *metabolomics*. According to J.J.Ramsdem [15], metabonomics is defined as "*the quantitative measurement of the multiparametric metabolic responses of living systems to patho-physiological stimuli or genetic modification, with particular emphasis on the elucidation of differences in population groups due to genetic modification, disease, and environmental (including nutritional) stress".* Although metabonomics usually compares profiles without identifying individual compounds, for the purpose of this chapter the two expressions will not be distinguished and the term metabolomics was elected to be used throughout.

Thus, the following paragraphs of this chapter will show the reader the benefits afforded by NMR metabolomics in improving the understanding, at the molecular level, of biological pathways involved in a variety of respiratory disorders.

CHOICE OF THE BIOLOGICAL FLUIDS TO BE ANALYZED

All biofluids that can be used for investigating the metabolomics of lung are characterized by a combination of strengths and weaknesses. Working on individuals with poor lung function, an important parameter that heavily influences the decision concerning which biofluid to sample is the patient's clinical status. In this respect, sampling procedures particularly safe and easy to perform would obviously be preferred over methods of collection with high degree of invasiveness, *i.e.* sampling of bronchoalveolar lavage fluid (BALf). BALf is collected by means of a fibro bronchoscope through which a liquid is instilled and fluid recovered by aspiration. Being rather invasive this procedure is practiced only when strictly necessary from a clinical point of view. Since for ethical reasons it cannot be applied to healthy volunteers, the analysis of healthy controls is prevented. Most likely due to the difficulties in sample collection, only few articles deal with the metabolic profiles of BALf [16 - 19]. By contrast, a good number of reports are focused on exhaled breath condensate (EBC), perhaps the source of information on lung status collected by the mildest processing. EBC consists in water vapor obtained by cooling, in commercially available equipments, exhaled air from spontaneous breathing. This procedure is completely non-invasive, allows collection of good amounts of EBC from patients and also from healthy controls without any discomfort or risk. The fact that EBC contains volatile and non-volatile compounds from the central airways, which reflect the composition of the airway-lining fluid, makes it a source of information complementary to BALf. As such, it could be a matrix potentially very helpful in metabolomic studies on lung disorders.

Another fluid collected by a mild procedure is sputum. When not spontaneously issued, it may be induced by inhalation of nebulized sterile hypertonic saline solution which induces the liquefaction of airway secretions. This treatment promotes coughing and allows patients to expectorate small amounts of sample quite easily [20]. While being currently associated to a very low risk, this process can become even less invasive if bronchodilators are added to the hypertonic solution. The wide variety of sputum components (*i.e.* cells and cellular debris resident within the tissue and/or the airways lumen, mucus, microbial products from any colonizing bacteria or viruses, particulate/inhaled matter derived from the environment), make it a wealth of information on lung conditions [21]. Due to the lack of recent articles on the metabolomics of sputum in the area of respiratory disorders, this fluid will not be discussed in this chapter.

Because also saliva is collected using non-invasive methods, it is potentially widely usable for metabolomic studies [22]. Unfortunately, the composition of salivary metabolome can be affected by many variables (spanning from diet to smoking, sampling time and even gender), that represent results-confounding factors [23].

The most commonly used matrices for metabolomics studies are plasma/serum whose sampling is low-invasive. The rationale for using these fluids as matrices for analyzing the profile of metabolites in pulmonary disorders is that, upon damage or death, lung cells release part of their content into the bloodstream. The considerable amount of different metabolites makes blood very informative of lung status. In fact, any change in the metabolite content that may indicate a pathological state is suitable for monitoring different disorders of the respiratory tract. However, the high complexity of blood makes its analysis extremely challenging and depletion of the large amounts of proteins required to perform metabolomic studies may, at least partially, invalidate the reliability of results [24].

The fact that urine is collected in a non-invasive and un-restricted way makes it available in large quantities. Although the high content of salts may represent a problem for NMR analysis, the use of urine for metabolomic studies may have significant clinical advantages: i) compared to other biofluids it contains relatively low concentrations of proteins and cells, ii) the concentration of metabolites is high [25], iii) requires minimal sample pre-treatment [26]. All these advantages are prompting the investigation of urine metabolomics to study pulmonary disorders.

The observation that the results from the analysis of plasma/serum, EBC and urine may be easily integrated motivated several authors, among those mentioned in this chapter, to carry on their metabolomic investigation simultaneously on two [27 - 31] or on a higher number of biofluids [32]. In most cases, aim of these studies was the unambiguous diagnosis of a specific lung disorder taking advantage of the combination of data generated by comprehensive analyses. In fact, in the presence of two or more disorders that have much in common in terms of clinical characteristics, patients are likely to go unrecognized if changes in the composition/level of metabolites in a single matrix do not provide a rationale for obtaining a differential diagnosis.

The experimental data generated with these studies emphasize the possibility of analyzing different matrices to diagnose a pulmonary disorder faster and more correctly through the detection of changes in the levels of metabolites detected in each matrix and crossing of data. This approach may also allow getting information about both localized and systemic changes.

The list of human fluids analyzed to investigate the metabolome of respiratory disorders considered in this chapter is shown in Table **1**. Additional information of the Table includes the level of invasion of fluid withdrawal, the level of standardization of method of collection, the main advantages and disadvantages of fluid under examination and the reference numbers of original articles.

Table 1. List of human fluids analyzed to investigate the metabolome of respiratory disorders considered in this chapter.

Human Body Fluid	Level of Invasivity of Fluid Withdrawl*	Level of Standardization of Method of Collection*	Advantages	Disadvantages	References
• Plasma/Serum	+	+++	-Low-invasive collection method -Very informative of lung status	-Deproteinization before analysis is mandatory -Systemic changes (may) introduce biases	[27, 28, 30, 62 - 64, 67 - 69, 73, 76, 83, 98, 107]
• Bronchoalveolar Lavage Fluid	+++	++	-Method of collection is well-standardized -Metabolites likely reflect the distress signals of the injured compartment	-Method of collection is very invasive -The fluid collected is generally diluted and should be concentrated for NMR analyses -High protein/salt content	[16 - 18]

(Table 1) cont.....

Human Body Fluid	Level of Invasivity of Fluid Withdrawl*	Level of Standardization of Method of Collection*	Advantages	Disadvantages	References
• Exhaled Breath Condensate	---	+/-	-Collection completely noninvasive -Contains both volatile and nonvolatile components	-Small volumes are currently collected from lung patients -The sample is diluted and must be concentrated for NMR analyses	[40, 43, 56, 57, 72, 75]
• Urine	---	+++	-Available in large/very large amounts -Collection completely non invasive -Stable composition -Relatively less complex than serum and plasma -Low concentration of total protein	-Lack of proximity and specificity to airway physiology -High salt content	[27, 28, 30, 31, 52, 66]
• Saliva	---	++	-Collected with a non-invasive method -Dilution may be normalized (amylase assay)	-Metabolome may be affected by several confounding factors	[22]

*+/- indicates the level of invasiveness and of standardization of the method. The higher the number of +, the higher the levels and vice versa

SAMPLE COLLECTION AND HANDLING

The procedures adopted for biofluid collection might have a great impact on the interpretation of metabolomic results being sometimes difficult to distinguish between pathophysiological responses and variations due to biological/technical issues. Adoption of standardized and reproducible procedures for sample collection and storage that guarantee the processing of all samples under the same conditions is thus a fundamental step. Uniformity in collection equipment and adoption of standardized collection protocols are crucial aspects that make the results of analyses reliable. Another critical point is the ability to "freeze" the clinical conditions of the individual under investigation at the time of biofluid collection. This may be achieved by keeping the composition of the biofluid as close as possible to the time point at which it was collected. For the storage, it is important to freeze the samples immediately after collection. By stopping the enzymatic activity, further transformations of metabolites can be avoided and the effects of factors that can cause changes in the metabolic profile minimized. However, like most biological molecules, metabolites degrade over time even when frozen. Thus, the earliest possible analysis of a sample and the avoidance of samples with freeze-thaw history prevent potential loss of biomarker analytes [33].

METHODS OF SAMPLE PREPARATION

Assessing and controlling the pre-analytical handling of a biological sample is fundamental for the reliability of downstream analyses that strongly depends on sample integrity. On turn, this is strictly connected with the availability of standardized protocols for their management. Hence, choosing the most appropriate sample preparation technique is often a daunting task. Nevertheless, (almost) all conventional (electrophoretic and/or chromatographic) protocols used to investigate body fluids include off-line or on-line purification steps as a prerequisite to eliminate undesired matrix constituents that can lead to impaired analyte detection. Operators may have to face two opposite situations: i) the massive removal of components that interfere with the separation system or ii) the mild removal of these substances. The first procedure may be detrimental for biospecimens that have affinity towards these clotting factors. Their fate would be to undergo an intolerable loss. Conversely, an insufficient sample clearing may result in a false analyte quantitation due to reduced specificity/accuracy of the assay. Thus, while preservation of analytes in the separation system is a major cornerstone in any sample preparation approach, the preferred route to prevent massive sample losses (or possible damage of sample integrity) does not follow a pre-established protocol. It seems plausible to conclude that the "ideal" case for obtaining the most reliable assessment of components in a biological sample would be avoiding any kind of pre-treatment/manipulation prior to their analysis. The major question is whether this ideal condition can actually be achieved when processing "real" samples. From a first glance to the articles considered in this chapter, the answer is certainly affirmative, the closest conditions to an ideal case being offered by NMR analysis. Having previously stated that one of the known advantages of NMR over other techniques is its ability to require a minimal sample pre-treatment (if any), this option does not appear such surprising. The experimental procedures described in these reports suggest that, despite working on different matrices, the sample preparation protocol is very simple and similar in all cases.

The protocol needed for plasma and serum samples is the most "labour intensive" because these samples must undergo a process to remove all proteins that would prevent NMR-based metabolomic studies. In few words, immediately after withdrawal, sample is frozen at -80 °C (to maximize the chemical stability of components) and thawed only at the moment of use. Proteins are removed by centrifugation (4,500-15,000 rpm) or by ultrafiltration through a 3kDa cut-off membrane. The supernatant or the filtrate are added of deuterated phosphate buffer to adjust the pH value and provide a necessary lock signal. The presence of the buffer is needed as chemical shifts are sensitive to pH changes that may determine a systematic bias in the data analysis. Trimethylsilyl propionate (TSP) is added to NMR samples as chemical shift reference.

The procedures indicated for urine and BALf suggest that these samples may be processed in a similar way. Samples are usually centrifuged (at 12,000-16,000 rpm) and aliquots of supernatant transferred to NMR tubes with the addition of a deuterated buffer (phosphate, NaCl or imidazole) and TSP or 4,4-dimethyl-4-silapentane-1-sulfonic acid (DSS) as chemical shift reference.

EBC and saliva are not subjected to pre-treatments. They are stored at -20 °C immediately after collection and added of D_2O and TSP or DSS (if needed) at the moment of use [22].

When plasma, urine and EBC are the matrices under investigation, an amount of NaN_3 (about 3 mM) is often added to the sample as a preservative to prevent bacterial growth.

A scheme that summarizes the simple procedures applied for the preparation of samples for NMR analyses presented in this chapter is shown in Fig. (**1**).

For more information on the design of metabolomics experiments in the field of respiratory diseases, the choice of the most appropriate matrix and the progress achieved by this and other omics technologies in the last ten years in this area, the reader is invited to refer to two excellent and comprehensive review articles here indicated [29, 34].

ACQUISITION OF SPECTRA AND DATA PROCESSING

The description of the mechanisms that govern NMR spectroscopy is outside the scope of this chapter. Readers who are not familiar with this technique and are interested in deepening the subject may find the basics of NMR spectroscopy in many excellent books, a few of which are indicated here [35, 36]. Aim of this paragraph is to provide insights into the specific procedures described in the articles that will be discussed in the following sections. The most important nuclei

Plasma/Serum

- Protein removal:
  Centrifugation (4500-15000 rpm)
  Ultrafiltration (3 kDa)
- + NaN_3 for plasma

Urine ⟶　　　Processed biofluid　　⟵ **BALf**

- Centrifugation (12000-16000 rpm)
- + NaN_3

- Centrifugation or not

+ Buffer (sodium phosphate, NaCl or imidazole)
+ D_2O
+ TSP or DSS

EBC ⟶　　Sample in NMR tube　　⟵ **Saliva**

+ D_2O (10%)
+ TSP or DSS
+ NaN_3

+ D_2O

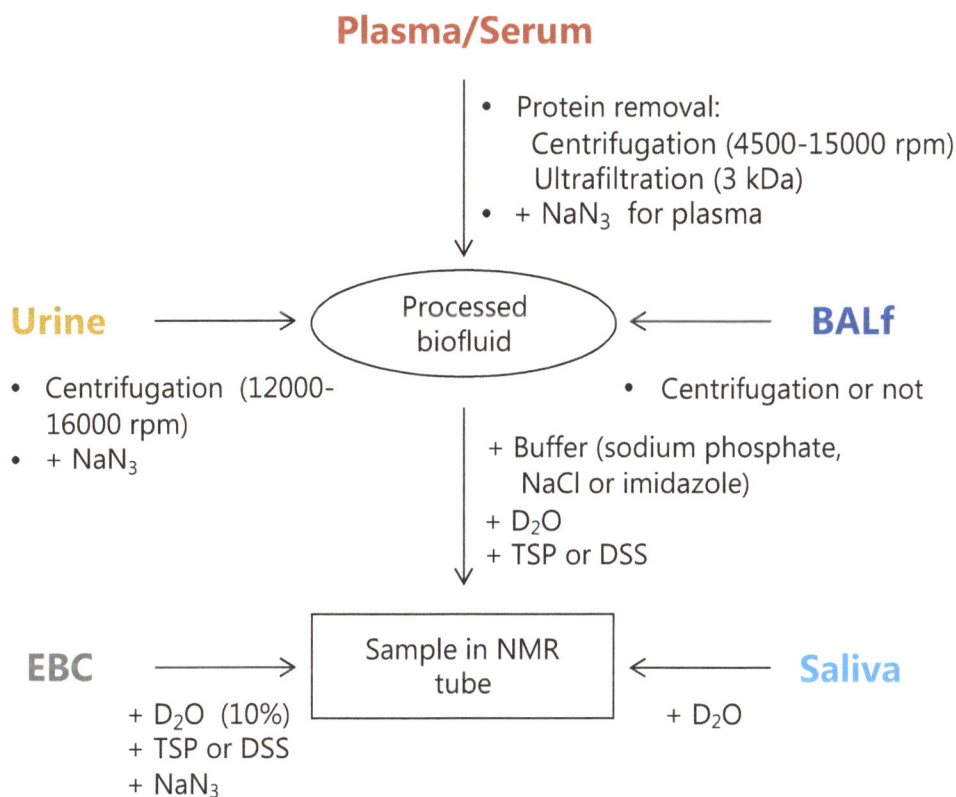

Fig. (1). Schematic depiction of the procedures applied, for each fluid, to prepare samples for NMR analysis.

in NMR studies of biological fluids are ^1H, ^{13}C, ^{15}N and ^{31}P. Due to its presence at near 100% natural abundance in samples analyzed, ^1H is the most sensitive and widely used nucleus in case-studies commented in this chapter and, more in general, in NMR spectroscopy. Two studies [37, 38] show that, although less frequently employed, also ^{13}C and ^{31}P are nuclei that can provide a plethora of information enabling researchers to distinguish between different cohorts of individuals analyzed. From a technical point of view, the NMR spectra of biological samples considered in these articles have been recorded applying external magnetic fields in the range between 400 and 800 MHz. More specifically, in two cases analyses were performed with equipments operating at 400 MHz [37, 38], in a single case at 500 MHz [22] and in two cases at 800 MHz [18, 19]. In the majority of cases here reported, spectra were recorded with instruments operating at 600/700 MHz.

The instrument field strength is well-known to have impact on line shape and sensitivity, the highest field providing the best sensitivity [39]. The screening of articles also evidenced that, in most cases, a simple mono dimensional (1D) ^{1}H NMR spectrum allowed the authors to collect detailed information on metabolic profile [39]. Metabolites were identified by spectroscopists either manually and/or with the assistance of software containing appropriate databases. In some cases, due to the very low level of metabolites in samples analyzed or to the presence of hundreds of signals which were partially or completely overlapped because of the similarity of their resonant frequencies, 1D-NMR was insufficient in providing a correct interpretation of data.

The authors have overcome these difficulties by applying two-dimensional (2D) NMR spectroscopic methods [40]. By including experiments such as ^{1}H-^{1}H correlated spectroscopy (COSY), in conjunction or not with ^{1}H–^{1}H total correlation spectroscopy (TOCSY) and ^{1}H–^{13}C heteronuclear single-quantum correlation (HSQC), these methods spread out the NMR signals across a second dimension and provided a tremendous improvement in signal resolution [19]. This allowed elucidating structural information on metabolites of interest with their consequent unambiguous identification even in the presence of very complex mixtures. As indicated above, since the design of the 2D NMR pulse sequences and the analysis of how these sequences provide the corresponding spectra are outside the scope of this chapter, they will not be discussed here. To get insights into these aspects, the reader is invited to refer to comprehensive books on this topic [41, 42].

Given the rationale of their research, most of reports discussed in this chapter were mainly focused on understanding whether NMR spectra from different cohorts behaved similarly (thus grouping in a single class), or differences in metabolic signatures could legitimate discrimination between groups. To extract relevant information from the large sets of data in their hand and discern meaningful patterns, authors have applied appropriate unsupervised and/or supervised multivariate analyses. Among the first, principal component analysis (PCA) and hierarchical clustering (HCA) [19, 30, 31, 43] were the procedures most commonly applied. Partial least squares (PLS) [19], partial least squares discriminant analysis (PLS-DA) and orthogonal partial least squares discriminant analysis (OPLS-DA) [19, 30, 43] were the most-used supervised approaches.

In general, the most applied method in articles considered is PCA *i.e.* an approach that uses a small set of variables, the so-called principal components. In this case, the dimensionality of data is reduced by clustering samples based on their inherent similarity/dissimilarity without any prior knowledge of class membership. By allowing an immediate observation of the variability in a data set and the identification of hidden phenomena, PCA provides an unbiased overview of differences/similarities among samples.

CASE STUDIES

The range of pulmonary disorders taken into consideration in this chapter testify the remarkable development of NMR techniques for the study of metabolomics in this area. The clinical cases investigated span from disorders that are the major cause of global morbidity and mortality, *i.e.* asthma, chronic obstructive pulmonary disease (COPD), cystic fibrosis (CF), pulmonary arterial hypertension (PAH), tuberculosis (TB), sarcoidosis, to diseases which are apparently less common/severe such as high altitude pulmonary edema, acute respiratory distress syndrome (ARDS), invasive pulmonary aspergillosis, pulmonary Langerhans cell histiocytosis (PLCH). Scrolling through these studies it can be inferred that metabolomics, while being still in its infancy compared to other omics technologies (transcriptomics, proteomics), can boast a good number of successes in the respiratory field. In most cases the task of this omic approach was to help unravel the complexity of a disease and enhance the interpretation of symptoms. To this purpose, the content of metabolites in both patients and healthy controls (or in patients submitted or not to a therapeutic intervention) was investigated in the search of qualitative/quantitative differences between cohorts. The rationale of these studies was the identification of sets of potential biomarkers of the disorder aimed at building a knowledge base of metabolite variation that could allow an improvement of disease diagnosis. Particular attention has been paid to the identification of specific metabolites/clusters of metabolites for the early and/or correct diagnosis of disease sub-phenotypes [44]. Given that existing criteria are often not able to provide this diagnosis, the metabolomic approach would play an important role for the characterization of distinct areas of metabolism and for drawing the biochemical pathways that characterize the disease phenotypes.

The application of metabolomics to the lung disorders indicated above resulted in the identification of classes of compounds that could lay the foundations for generating hypotheses on these pathologies. These findings have been summarized in Table **2**. The table also includes additional information *i.e.* the tentative number of metabolites identified in the different fluids analyzed for each disorder considered; the chemical nature of these compounds and the pathways they are potentially involved in.

The most important metabolites and metabolic pathways that characterize the difference in metabolic profiles between healthy controls and patients with respiratory disorders will be discussed in the following paragraphs.

Taurine/Hypotaurine Metabolism

From among all metabolites differentially expressed in biofluids above mentioned, taurine is certainly one of the most important. Despite it lacks the carboxyl group and is not used in protein synthesis, taurine (β-aminoethane sulfonic acid) is often referred to as an amino acid. However, given the variety of biochemical and physiological roles played by taurine [45], its broad distribution in the human body and the wide range of conditions (that span from the central

Table 2. Chemical nature of metabolites identified in the different fluids analyzed for each disorder considered and the pathways they are potentially involved in.

Disorder	Human Fluid Analyzed	Approximate Number of Perturbed Metabolites Identified	Chemical Nature of Metabolites	Metabolic Pathways Potentially Involved	References
Asthima	Serum EBC Urine	40	Amino acids Krebs cycle intermediates Keto and Hydroxyacids Carbohydrates Nucleosides Nucleotides Peptides Fatty acids	Pyruvate pathway Krebs cycle Gly, Ser and Thr pathway Arg and Pro metabolism Taurine and hypotaurine pathway Glyoxylate and dicarboxylate pathway His pathway	[31, 40, 52, 62, 66]
COPD	Urine EBC Plasma/ Serum	50	Amino acids and amino acid precursors Krebs cycle intermediates Fatty acids Carbohydrates Keto and Hydroxyacids Alcohols Ketone bodies Triglycerides	Krebs cycle Taurine and hypotaurine pathway Pyruvate pathway Val, Ile and Leu pathway His pathway Arg and Pro metabolism Glutamate/glutamine metabolism Lipid catabolic pathways Bacterial catabolic pathways Niacine metabolsim	[27, 28, 30, 43, 52, 72, 56, 57, 63, 67, 73, 98]
Cystic Fibrosis	EBC BALf	10	Alcohols Amino acids Keto acids	Taurine and hypotaurine pathway Pyruvate pathway Lipid catabolic pathways Bacterial catabolic pathways	[17, 75]

(Table 2) cont.....

Disorder	Human Fluid Analyzed	Approximate Number of Perturbed Metabolites Identified	Chemical Nature of Metabolites	Metabolic Pathways Potentially Involved	References
Pulmonary Arterial Hypertension	Plasma	20	Amino acids Krebs cycle intermediates Keto and hydroxyacids Carbohydrates Lipids	Krebs cycle Val, Ile and Leu pathway His pathway Glutamate/glutamine metabolism Lipid catabolic pathways Bacterial catabolic pathways	[76]
Sarcoidosis	Saliva Serum	12	Alcohols Amino acids Krebs cycle intermediates	Krebs cycle Pyruvate pathway Lipid catabolic pathways	[22]
Tuberculosis	Plasma/ Serum	30	Krebs cycle intermediates	Pyruvate pathway Krebs cycle His pathway Glutamate/glutamine metabolism Bacterial catabolic pathways	[64, 69, 83]
High-Altitude Pulmonary Edema	Plasma	15	Amino acids Phospholipids Carbohydrates	Glutamate/glutamine metabolism Lipid catabolic pathway	[107]
Acute Respiratory Distress Syndrome	BALf	6	Amino acids Keto acids	Lysine metabolism Taurine and hypotaurine pathway Arg and Pro metabolism Gly, Ser and Thr pathway Glutamate/glutamine metabolism	[18]
Bronchiolitis Obliterans Syndrome	BALf	15	Amino acids Keto and Hydroxyacids Carbohydrates	Glutamate/glutamine metabolism Taurine and hypotaurine pathway Lipid catabolic pathways Pyruvate metabolism	[16]

nervous system, to cytoprotection, cardiomyopathy, renal dysfunction and severe damage to retinal neurons) in which it was shown to be beneficial, the definition of "very essential" would be much more appropriate. Adult mammals may synthesize taurine in the liver and, to some extent, in central nervous system

(CNS), lungs, skeletal muscle, adipose tissue and mammary glands [46]. The synthesis occurs through the two different metabolic pathways shown in Fig. (**2**). In the first one cysteine is converted in cysteine sulfinate and hypotaurine *via* the sequential action of cysteine dioxygenase (CDO) and cysteine sulfinate decarboxylase (CSAD). Hypotaurine then undergoes oxidation in a reaction that follows a still unknown mechanism. A minor synthesis pathway involves the breakdown of Coenzyme A with the production of cysteamine followed by oxidation to hypotaurine by the enzyme cysteamine dioxygenase (ADO) [46].

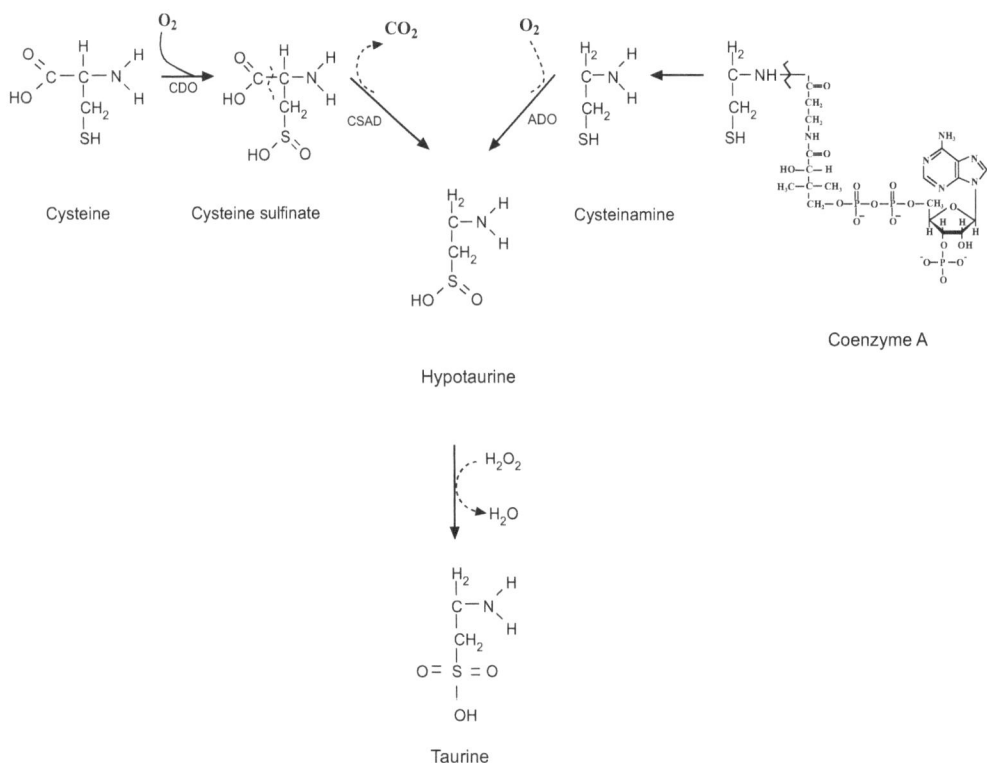

Fig. (2). Scheme showing the possible pathways for the synthesis of hypotaurine and taurine.

Besides being involved in the regulation of cell volume and in the stabilization of cell membrane, taurine has also anti-oxidative, anti-inflammatory and anti-apoptotic effects [47 - 49]. Furthermore, taurine is known to act as a weak agonist for chloride-permeable gamma-aminobutyric acid type A receptors (GABA$_A$R) and glycine receptors (GlyR). These receptors are present not only in the neural synapsis but also in the CNS and in the lungs [50]. Under normal conditions, the binding of taurine to the α_4-subunit of GABA$_A$R and to GlyR induces the increase of airway smooth muscle cells relaxation and stimulates mucus secretion from goblet cells. As previously reported, another important role of this amino acid is the regulation of cell volume that occurs through its accumulation *via* the Na^+-dependent Tau-T and the proton coupled PAT-1 transporters and its release through volume-insensitive and volume-sensitive leak pathways (VSOAC) [46, 51]. The activity of the transporter Tau-T is deeply down-regulated by acidification, osmotic cell swelling and presence of reactive oxygen species (ROS). At the same time, the effect of volume sensitive tyrosine kinases (PTK) is amplified by ROS through the inhibition of specific phosphatases and kinases, thus prolonging the VSOAC opening, which is also up-regulated by eicosanoids released in response to inflammation [46]. An increase of taurine concentration in several biological fluids has been reported in various types of airway diseases including COPD, asthma, CF, ARDS and Bronchiolitis Obliterans Syndrome (BOS) [16 - 18, 31, 52]. Since taurine accumulates intracellularly in airway epithelia regulating the osmotic balance, its increase in samples of patients affected by lung diseases with different levels of severity could reflect epithelial cell damage or inflammation-induced osmotic stress. In addition, taurine accumulates in the cytoplasm of neutrophils and other leukocytes and its presence may reflect the presence of these cells in the inflamed airways [17]. On the other hand, the increase of mucus secretion during inflammation is triggered by the release of ROS by neutrophils and macrophages and of eicosanoids by leukocytes that, in turn, stimulates a decrease of the intracellular taurine intake and an increase of its release in extracellular fluids. The persistence of the inflammatory state causes a worsening of lung conditions due to autocrine stimulation of cysteinyl leukotriene receptor 1 (CysLT1) induced by leukotriene D_4. CysLT1 is a strong constricting agent and counteracts the relaxing effect of taurine on smooth muscle cells.

The increase in taurine levels during the pathogenesis of ARDS is an index of rescue mechanism to control inflammation. In fact, its protective role as an antioxidant with anti-inflammatory properties helps in reducing the free radical production. Moreover, taurine mitigates IL-2 induced lung injury by attenuating neutrophil-endothelial interaction [18].

Metabolism of Short Chain (C_2 to C_4) Compounds

As shown in Fig. (**3**), a series of short-chain (C_2 to C_4) biological compounds belonging to different chemical subclasses and strictly connected to each other in terms of molecular structure have been identified in biological fluids of patients affected by different lung diseases.

It must be underlined that the detection of many of these compounds (ethanol, butanol, 2-propanol, acetoin, 2,3-butanediol), that are products of bacterial fermentation, unambiguously indicates the presence of microbial infections in lungs of affected patients. Bacterial "colonization" of the airways is quite perilous as it is always associated to high levels of inflammation, increased exacerbations and accelerated decline in lung function [53]. Bacteria are known to produce several volatile compounds (alcohols and ketones) which derive from fatty acid degradation and that could be used as biological markers for the presence of pathogens.

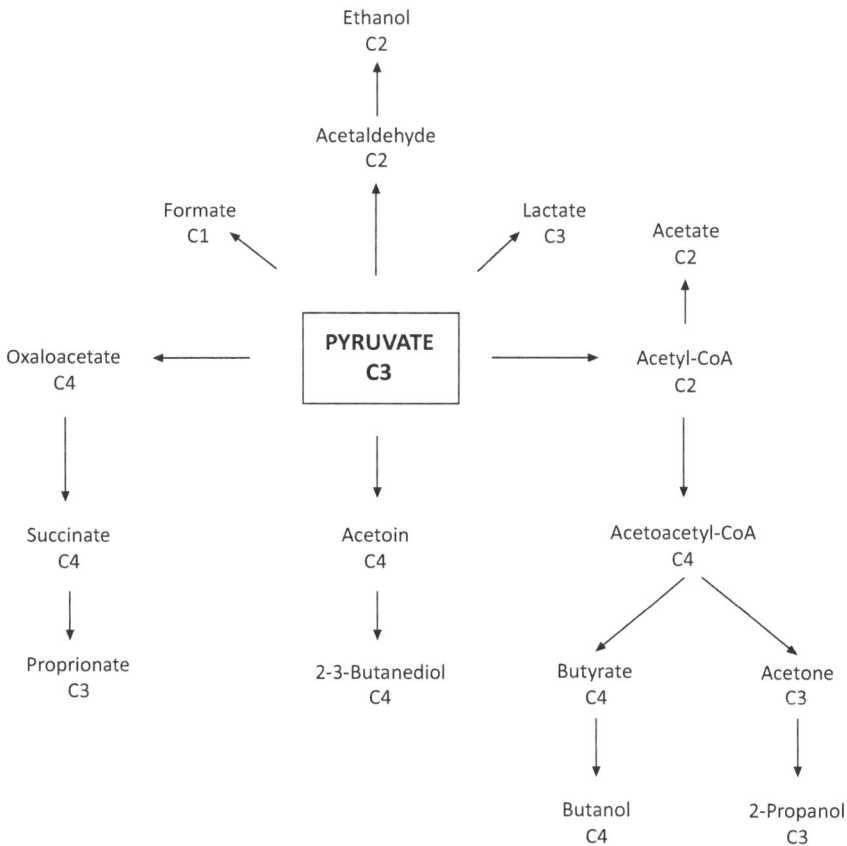

Fig. (3). Relationships between the short-chain metabolites identified in biological fluids.

From among these compounds, acetoin (also known as 3-hydroxybutanone) is undoubtedly worthy of mention. It is a neutral, four-carbon molecule used by several fermentative bacteria as an external energy store. Acetoin is converted into acetyl-CoA through a reaction catalyzed by the acetoin dehydrogenase complex following a mechanism similar to that of the pyruvate dehydrogenase complex. However, differently from this latter, acetoin does not undergo decarboxylation by the E1 enzyme and, as shown in the scheme of Fig. (**4**), its catabolism results in the release of a molecule of acetaldehyde [54]. Owing to its neutral nature, acetoin produced and excreted during exponential growth prevents over-acidification of the cytoplasm and of the surrounding medium that would result from accumulation of acidic metabolic products, such as acetic and citric acids [55].

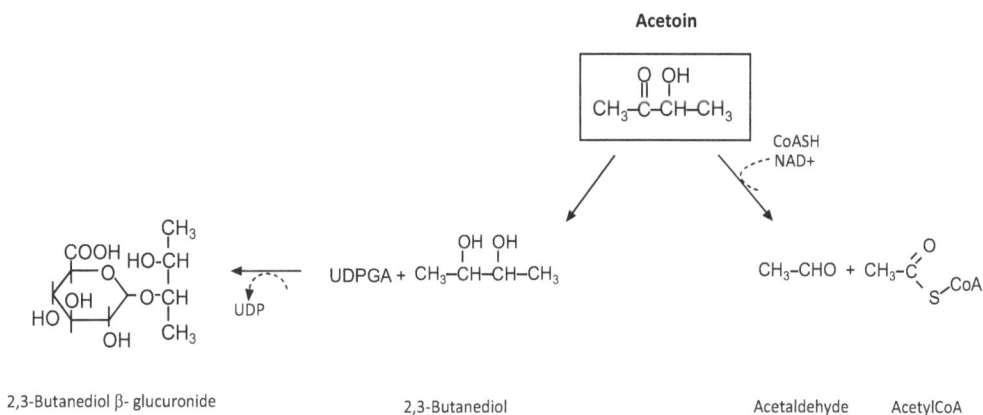

Fig. (4). Catabolic pathway of acetoin.

The finding in EBCs of patients affected by alpha1-antitrypsin deficiency (AATD) of acetoin levels 150-fold higher than those found in healthy controls [56] allowed to hypothesize that the concentration of this metabolite could be a good parameter for discriminating the two cohorts under investigation. According to de Laurentiis *et al* [43], who investigated the metabolome of EBCs from patients with COPD and PLCH, acetoin could help discriminating between these two groups and, in turn, between both groups of patients and healthy smokers. In some bacteria, acetoin is reduced (by acetoin reductase/2,3-butanediol dehydrogenase) to 2,3-butanediol that is coupled to UDP-glucuronide forming 2,3-butanediol β-glucuronide to be excreted into urine (Fig. **3**). Thus, it can be speculated that detection of this compound may be a way for checking the bacterial colonization of lungs [43, 56].

Propionate, a short-chain fatty acid, was found to be increased in EBC of patients affected by AATD, COPD and PLCH compared to controls [43, 56, 57]. This compound is assumed to have an anti-inflammatory role. Short-chain fatty acids are known to regulate a series of leukocyte functions including production of cytokines, eicosanoids and chemokines and, most likely, also affect leukocyte migration to the loci of inflammation [58]. Given the role of propionate in the inhibition of cholesterol synthesis, an involvement of lipid metabolism in this disorder may be suggested also in the light of the higher lipolysis associated with smoking that leads to elevated serum levels of free fatty acids [43]. This hypothesis was also supported by the contemporary increase of acetate and butyrate observed in the above patients [43, 56].

The correlation between the values of propionate and those of prostaglandin E_2 (PGE_2) observed in COPD patients suggests that the levels of the former analyte may reflect respiratory inflammation [57].

Formate is the only non-tetrahydrofolate(THF)-linked intermediate in one-carbon metabolism: it is produced in mammals from a variety of metabolic sources and occurs in serum of adults (at a concentration of approximately 30 µM). Formate is a source of one-carbon groups for the synthesis of 10-formyl-THF and other one-carbon intermediates that are used for purine and thymidylate synthesis, and the provision of methyl groups for synthetic, regulatory, and epigenetic methylation reactions. Even though it is largely produced in mitochondria, these functions mostly occur in the cytoplasm and nucleus.

An increased methylation of arginine is thought to be a key process related to the pathogenesis of asthma [59, 60]. About this, increased levels of methylarginine (monomethylarginine and asymmetric dimethylarginine) have been described in patients with severe asthma compared to subjects with non-severe disorder [59]. In both allergic diseases and asthma, arginine methylation was also observed to regulate cytokine over-expression [60]. Furthermore, several studies have suggested that DNA methylation alters immune responses towards T-helper cell type 2 (Th2) asthma phenotype enhancing the risk of airway inflammation [61, 62]. A significant decrease of metabolites related to the enhancement of the methyl transfer pathway, including formate, choline, O-phosphocholine, and methanol was observed in patients with asthma [40, 62] and a significant correlation was found between arginine levels and the predicted values of FEV1% [62]. Decreased levels in serum of asthma patients of O-phosphocholine, a component of the endothelial cell barrier and a pulmonary surfactant, could indicate a lack of protection of the alveolar region and of the conducting airways [62].

The demonstration that formate levels detected in COPD patients are correlated to PGE_2 suggests that this metabolite may reflect respiratory inflammation [57, 63]. Conversely, since formate is also produced by gut microbioma, urinary concentration of this compound seems to be indicative of variable gut microbiome composition among individuals with different lung function [28].

The reduced serum levels of formate in TB patients might reflect an increased requirement for nucleotide biosynthesis in an inflammatory condition and the increase in nucleotide metabolism after *M. tuberculosis* (MTB) infection probably reflects active host inflammatory cell division [64].

Acetoacetate, β-hydroxybutyrate, and their spontaneous breakdown product, acetone, are three water-soluble molecules (also known as ketone bodies) produced by the liver from fatty acids as a result of intense gluconeogenesis during fasting; starving; low carbohydrate diets; prolonged exercise or untreated type 1 diabetes mellitus [65]. These molecules are readily utilized by the extra-hepatic tissues and converted into acetyl-CoA that enters the citric acid cycle and is oxidized in the mitochondria for energy production. In the brain, ketone bodies are also used to produce acetyl-CoA involved in the long-chain fatty acid synthesis.

Increased levels of ketone bodies have been observed in serum and/or urine from asthma, COPD and TB patients [64, 66 - 69]. This increase suggests the use from these patients of storage lipids as an alternative energy resource. Persisting airway inflammation together with irreversible airflow obstruction and systemic inflammation that leads to consumption of body energy and malnutrition are indeed responsible for the disorganized lipid and amino acid metabolism [67, 68]. In fasting and starving states, an increase in short-term levels of branched-chain amino acids (BCAAs) occurs in parallel with increased muscle protein degradation [68, 70]. After several weeks, the rate of this degradation decreases together with blood concentration of BCAAs, whereas ketone body level increases. In COPD patients with emphysema, a decrease in serum creatine and its precursors, glycine, guanidine acetate and dimethylglycine has also been observed [68]. This finding suggests a reduced mitochondrial function, a hypothesis that is supported by the observation, in the early stages of COPD, of a block of electron transport chain coupled to excessive production of reactive oxygen species in skeletal muscle mitochondria [68, 71]. The increased acetoacetate and acetone levels linked to decreased levels of LDL observed in TB patients are also consistent with enhanced lipid degradation [64, 69].

An increase in acetate (mainly utilized by organisms in the form of acetyl-CoA) has been observed in EBC and/or serum/plasma of COPD, CF, BOS and PAH patients [30, 43, 56, 57, 72 - 76] and in saliva of sarcoidosis patients [22]. This increase might be due to the acceleration of lipid catabolism needed to satisfy the energy requirements caused by the typical poor nutritional status of these patients [30, 43, 74]. The possible correlation between the increased concentration in acetate and the energy needs is supported by the flux of exogenous acetate to acetylcarnitine observed in C57BL/6 mice [77]. The carnitine system is, in fact, mainly involved in the transport into mitochondria of fatty acids that are converted in energy [43, 77]. In line with the rise in acetate, also the propionate increase suggests the involvement of lipid metabolism, given that its production seems to inhibit cholesterol synthesis, leading to elevated serum levels of free fatty acids [78]. The elevated levels of acetate in COPD patients may also suggest acetylation of pro-inflammatory proteins in the airway lining fluid [72].

2-Propanol, an enzyme-mediated product of acetone reduction, was found to be increased in COPD patients [43]. Endogenous 2-propanol derives from the reduction of acetone by liver alcohol dehydrogenase, mainly when high levels of acetone and high $NADH/NAD^+$ ratios occur, as observed in ketosis [43]. The increase in concentration of 2-propanol, found in EBC from CF patients infected with *Pseudomonas aeruginosa*, was supposed to be due to bacterial metabolism and/or increased lipolysis and lipid peroxidation [74].

TCA Cycle Intermediates

The tricarboxylic acid (TCA) cycle consists in a series of chemical reactions used by all aerobic organisms to release stored energy into adenosine triphosphate (ATP) through the oxidation of acetyl-CoA (derived from carbohydrates, fats and proteins) and to provide the reducing agents NADH and $FADH_2$ as well as the precursors of certain amino acids. Interestingly, the level of a good number of metabolites involved in this metabolic pathway (succinate, fumarate, oxaloacetate, cis-aconitate and 2-oxaloglutarate) was found to be higher in urine of asthma patients who had recently suffered an exacerbation, compared to healthy controls [66]. This up-regulation of the TCA cycle could make one think of a greater effort of patients to breath during exacerbation and/or to the hypoxic stress resulting from the poor oxygenation occurring during exacerbation. The confirmation that an over expression of TCA metabolites is a consequence of the hypoxic stress occurring during exacerbation has come from a similar up-regulation in the TCA cycle observed in individuals during physical exercise [66].

Besides its role as an intermediate of the TCA cycle, succinate may also be considered an inflammatory signal since it directly regulates hypoxia-inducible factor-1α (HIF-1α) signaling thus promoting inflammation by macrophages [67]. The levels of succinate were found decreased in sarcoidosis patients compared to healthy controls although the levels of other TCA intermediates were not significantly different in the two groups [79]. In response to Toll-like receptor 4 (TLR4) activation, the macrophage metabolism is known to be switched from oxidative phosphorylation to glycolysis. This switch leads to the rapid accumulation of succinate that is also generated from isoleucine and glutamine that "feed" the TCA cycle [80]. Thus, the decrease of succinate, associated to a decrease in glutamine and isoleucine levels in sarcoidosis patients, can be a strong indication of a disturbance in the amino acid and energy metabolism [80]. This could provide a connection with the overall decreased energy levels and elevated fatigue observed at clinical diagnosis [81].

The normal levels of other TCA intermediates observed in sarcoidosis patients could be explained by the role played by citrate in the metabolic switch that regulates mitochondrial dysfunction during hypoxia or inflammation [80]. In fact, citrate can be converted into cytosolic acetyl-CoA that can be used as a precursor for fatty acid synthesis through an alternative pathway involving the citrate shuttle without passing through the conventional clockwise steps of the TCA cycle [82]. This process may contribute to the maintenance of other TCA intermediates (malate, oxaloacetate and citrate) at normal levels.

The state of hypoxemic stress of lung tissue in pulmonary disorders is also confirmed by the apparent decrease of succinate levels in COPD patients and by the mitochondrial abnormalities observed in their muscles [52].

The decrease of citrate level in urine of patients affected by COPD is due to respiratory failure which reflects the high-demand metabolic state [27]. On the other hand, COPD patients undergoing administration of doxycycline, a broad-spectrum antibiotic belonging to the tetracycline class that has been demonstrated to decrease this activity, show an increase in the level of this metabolite [63]. This increase is supposed to be due to the reduced activity of ATP-citrate lyase (ACL) [63], an enzyme responsible for catalyzing the conversion of citrate and CoA into acetyl-CoA (that can be used for fatty acid synthesis) and oxaloacetate along with the hydrolysis of ATP [63].

The decrease in the amount of serum fatty acids observed in these patients is a further support of this hypothesis.

The TCA intermediates found to be altered in the studies mentioned above are shown in the scheme of Fig. (**5**).

Pyruvate/Lactate Metabolism

Pyruvate is a key intermediate in several metabolic pathways. It derives from glucose through glycolysis, and is converted to a variety of compounds that, in addition to those previously shown in Fig. (**3**), include glucose itself *via* gluconeogenesis.

Given the role of pyruvate to supply energy to cells through the TCA cycle in the presence of oxygen, changes in serum pyruvate levels may suggest multiple metabolic dysregulation. The key enzymes regulating pyruvate metabolism are pyruvate dehydrogenase and pyruvate carboxylase by which pyruvate is transformed into acetyl-CoA and oxaloacetate, respectively. Both compounds can be shuttled through the TCA cycle. As previously reported, the reduction of pyruvate in human cells generates L-lactate. Since this reaction does not require oxygen, increased levels of lactate have been usually reported during periods of anaerobic exercise [34].

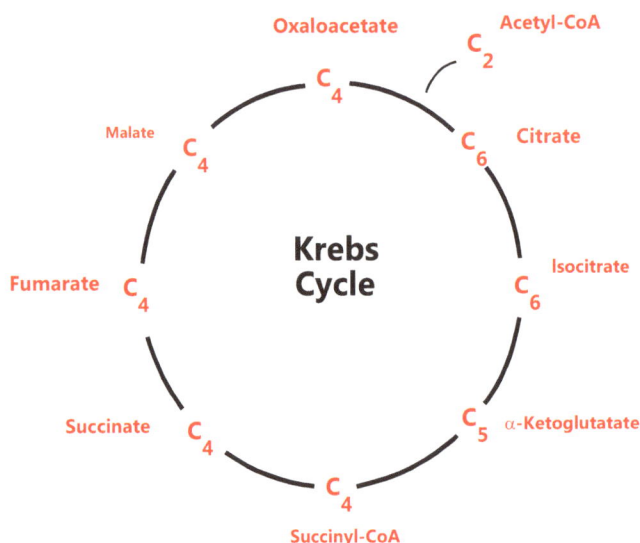

Fig. (5). The Krebs' cycle metabolites found altered in pulmonary disorders are indicated in bold.

The significant increase in pyruvate levels observed from the metabolomic analysis of serum from sarcoidosis patients suggested the activation of the glycolytic pathway in these subjects [79]. An increase of pyruvate and lactate was also observed in serum and plasma from TB patients [64, 69, 83]. It has been demonstrated that these patients experience malnutrition, weight loss and metabolic disorders [84]. The increased levels of pyruvate mentioned above suggest increased catabolism of all three major nutrients as well as increased energy consumption, a condition similar to metabolic changes observed in patients during tumor development [64, 69, 85]. In addition, increased levels of pyruvate and lactate are consistent with increased anaerobic glycolysis resulting from the granulomatous inflammation in the lung with central necrosis and tissue hypoxia induced by MTB infection [64, 69]. It has been speculated that an increase in lactate levels is associated with the progression of malignancy through the formation of tumor necrosis [86]. In fact, in malignant tissues, pyruvate is mostly converted to lactate and energy is produced anaerobically through the Warburg effect, even when the amount of oxygen is enough to support the mitochondrial function [69]. Therefore, the accumulation of lactate in TB patients could be considered as an index of tissue hypoxia and extent of necrosis with the infection progression. In addition, the fact that more pyruvate is converted into lactate, rather than entering the TCA cycle pathway, is probably the result of insufficient oxygen supply occurring during MTB infection [69].

The increase in pyruvate concentration could also occur through the methylcitrate cycle (MCC) in which propionyl-CoA is oxidized to pyruvate. In several bacteria and fungi, including *Mycobacterium tuberculosis*, MCC is involved in the assimilation of propionyl-CoA generated by β-oxidation of odd-chain fatty acids and by the branched-chain amino acids catabolism [64, 87]. Furthermore, MCC is also required for the detoxification of propionate [64, 87].

Also the metabolic activity of activated T-cells may contribute to the increase of lactate level. In fact, it has been reported that activated T cells use glycolysis to gain energy, being this process advantageous for the cell as it provides ATP more rapidly than oxidative phosphorylation [69, 88]. Furthermore, cytokine secretion is supported by the shift to glycolysis in lymphocytes. In fact, since the T-cells metabolism is disturbed by the intracellular accumulation of lactic acid, they secrete this compound with high efficiency [69, 88].

Lactic acidosis is commonly observed in patients with acute severe asthma. A plasma lactate concentration greater than 2.2 mM has been reported in 83% of children affected by this disease [89]. This increase is caused by different pathogenic mechanisms, including administration of β2-adrenergic agents, hypoxemia and respiratory muscle activity. The mechanism underlying β2-agonists-induced lactic acidosis still remains uncertain [90]. The high levels of lactate evidenced from the analysis of urinary metabolome of patients affected by asthma exacerbation, further support the hypothesis that these individuals were undergoing hypoxic stress [66]. On the other hand, as reported above [34], increased levels of lactate have been observed during anaerobic exercise even though the increase in concentration of lactate, during these periods, corresponds to a decrease in the abundance of citrate [91] which has not been observed in asthma patients. Lactic acidosis originated from increased lactate levels produces, in turn, renal tubular acidosis leading to a decrease in urinary levels of glycine and hippurate [34, 92] whose role as biomarkers of reversible renal function has already been described [93]. The lower hippurate levels observed also in EBCs from mild asthmatic patients seem to confirm this hypothesis [34]. However, the report of normal amounts of these two latter metabolites in urine of exacerbated asthma patients [66] suggests that their hypoxic stress is not sufficiently severe to cause renal malfunction. Increased levels of lactate in asthma patients are often associated with an increase in alanine concentration as part of gluconeogenesis [66].

Respiratory muscle fatigue causes an increase of lactic acid leading to a state of lactic acidosis that has also been reported in patients affected by COPD [30, 57, 63, 67, 73]. Given that high lactate levels are associated with restricted exercise capacity of COPD patients [83], this condition can complicate clinical evaluation and treatment in COPD [94]. It has been demonstrated that the administration of a specific pharmacological therapy to patients affected by COPD causes a significant decrease in serum lactate levels. The significant decrease of lactate in add-on doxycycline patients compared to controls underlined the effect of this therapy. This result was supported by the biochemical estimation of serum lactate which indicates a significant decrease in add-on doxycycline group compared with the pre-treatment group. A significant decrease of lactate after treatment with inhaled budesonide/formoterol of COPD patients affected by emphysema without and with bronchial wall thickening (phenotype E and M, respectively) demonstrates that anaerobic glycolysis energy supply is reduced after treatment with bronchodilators and corticosteroids [67].

Also the metabolic profile achieved from BALf of CF pediatric patients with high inflammation shows levels of lactate that reflect the hypoxic environment present in the infected CF lung [17]. Since deepest layers of infected airway biofilms are characterized by low oxygen concentrations [95], the airway epithelia located in these deepest layers are probably forced to rely on anaerobic metabolism for energy. This obviously results in increased lactate concentration. The high amounts of lactate found in plasma of patients with pulmonary arterial hypertension and in saliva of patients with sarcoidosis reflect the hypoxic state that characterize these diseases leading to glucose hyper-metabolism [22].

Amino Acid Metabolism

BCAAs Metabolism

Branched-chain amino acids (BCAAs) have several metabolic and physiologic roles. From a metabolic point of view, they promote protein synthesis and turnover, signaling pathways and metabolism of glucose [96, 97]. Oxidation of BCAAs may increase fatty acid oxidation, thus playing a role in obesity. From a physiological point of view, BCAAs have been demonstrated to fill roles in the immune system and in brain function as well. In fact, they are required for lymphocyte growth and proliferation and cytotoxic T-lymphocyte activity [96]. Lastly, BCAAs share with aromatic amino acids the same transport protein into the brain where they seem to play a key role in protein and neurotransmitter synthesis and production of energy [96].

The BCAAs metabolism is the most significant area of metabolism that helps discriminating COPD patients from healthy controls [68, 73]. In fact, valine, isoleucine, and their degradation product 3-hydroxy isobutyrate are usually reported at lower values in COPD patients (especially those suffering from severe and very severe COPD) compared to healthy controls [30, 34, 67, 68, 72, 73, 98]. This change in BCAAs metabolism could be the result of the typical cachexia observed in these patients. In the presence of extended periods of fasting, the degradation of skeletal muscle by proteolytic enzymes and the transamination of BCAAs by branched-chain aminotransferase (BCKHD) provide a resource for gluconeogenesis that is not suppressed by glucose in patients suffering from cachectic weight loss [34]. This consideration is supported by the finding that the reduction of BCAAs levels in COPD patients is correlated with the body mass index (BMI) [68].

On the other hand, the increase in BCAA catabolism, observed in urine of individuals after exercise [91], was reasonably attributed to proteolysis of the skeletal muscle and visceral region [34] as a result of increased gluconeogenesis in response to increased demand for energy during exercise [67, 68, 72, 73]. If this conclusion proves true, the lower levels of BCAAs in COPD patients could be the consequence of muscle proteolysis, potentially due to wasting. The increase of other essential amino acids, including phenylalanine, are consistent with increased protein degradation.

An increase in BCAAs has also been observed in plasma of patients with pulmonary arterial hypertension [76]. Notwithstanding that high blood BCAA concentrations are thought to be responsible for the oxidative damage at the glial and neuronal levels [99], the detrimental endothelial effect of this increase in BCAA levels is not known [76].

Histidine Metabolism

Being the precursor of several biologically relevant molecules, histidine is critical in many enzymatic reactions, some of which are schematized in Fig. (**6**).

Through a reaction catalyzed by a specific decarboxylase, histidine originates histamine, a molecule that contributes to inflammatory responses and constriction of smooth muscle [62, 100] whose levels were found to be increased in serum from asthma patients [52, 62]. Serum levels of 1-methylhistamine, a downstream metabolite of histamine, have been demonstrated to increase in asthmatic patients compared to healthy controls, especially after an asthma attack [52]. Conversely, these levels are decreased after administration of antiallergy medications [52, 101]. Also the histidine levels are higher in asthmatic patients compared to healthy controls [52]. Since histidine is negatively associated with inflammation and oxidative stress, it has anti-inflammatory properties [27]. This is consistent with the reduced levels observed in COPD patients and the negative correlation of this amino acid with radiographic emphysema [27]. Conversely, the levels of histidine increased in patients with pulmonary arterial hypertension (PAH) associated to systemic sclerosis (SS) [76]. It could be speculated that this increase is promoted by the anti-inflammatory effects of this amino acid on human coronary arteries. This activity is thought to be mediated by inhibition of interleukin-6, CD62E, and nuclear factor κB, which are involved in the pathogenesis of atherosclerosis and vasculitis [102]. Another intermediate of histidine catabolism is urocanic acid that originates from histidine by an ammonia lyase. Interestingly the levels of this compound in asthma patients were lower than in healthy controls [40]. Since the histidine-histamine conversion is favoured over the histidine-urocanic acid pathway, it has been hypothesized that monitoring the histidine conversion pathways could become a way of controlling asthma severity [40].

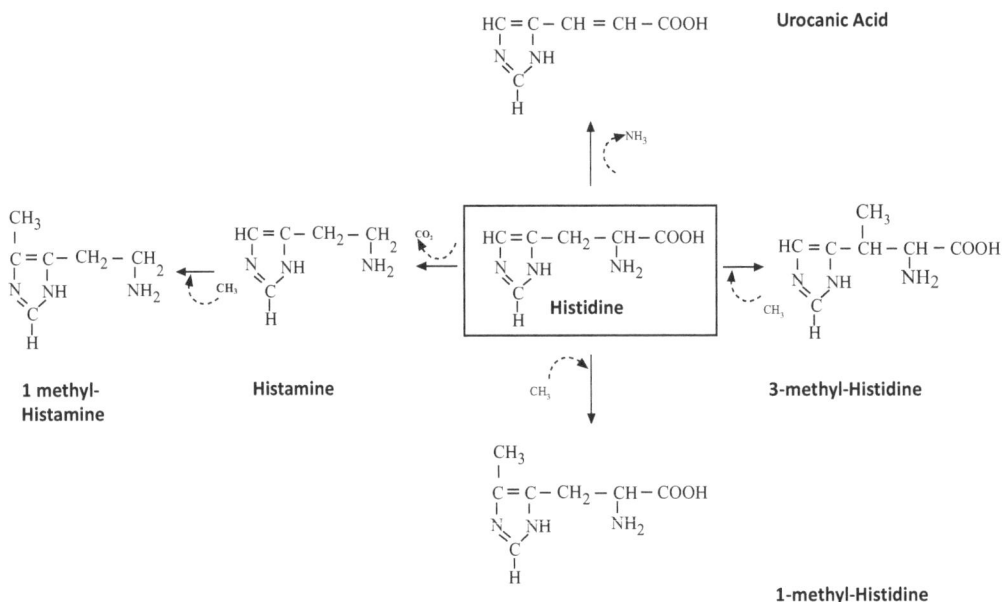

Fig. (6). Metabolites derived from histidine catabolism.

The conversion of histidine to 3-methylhistidine during the synthesis and crosslinking of muscle proteins may serve as a typical biomarker for skeletal muscle damage [68]. The increased concentrations of this amino acid observed in serum of emphysematous COPD patients suggest an increased degradation of muscle actin and myosin even in not cachectic patients [68]. The lack of correlation between elevated levels of serum 3-methylhistidine and either BMI or fat free mass (FFM) in patients with emphysema suggested that increased muscle protein turnover is a feature of COPD related to emphysema that precedes or is unrelated to the development of cachexia [68].

Also the increase of 1-methylhistidine in serum of TB patients may reflect accelerated muscle protein degradation. Vitamin E deficiency that leads to increased oxidative effects in skeletal muscle may also be responsible for 1-methylhistidinuria [64]. This is consistent with malnutrition and wasting observed in TB patients [64].

Arginine Metabolism

Besides its role in protein synthesis, arginine (a semi-essential amino acid), is also an important substrate for nitric oxide synthases (NOS) and arginases induced simultaneously in inflammatory conditions [59]. Arginine can be converted to citrulline and nitric oxide (NO) or to ornithine and urea through enzymatic reactions catalyzed by NOS and arginases, respectively [59, 103]. Expression of type 2 NOS in airways is induced by proinflammatory cytokines (*i.e.* interleukin-1, tumor necrosis factor α and interferon γ) and NO concentration was shown to be increased in EBCs from asthma subjects compared to healthy controls [59]. Moreover, patients presenting acute asthma exacerbations showed an increase in arginase activity and decreased levels of arginine in urine and serum [52, 59, 104]. Enhanced arginase activity coupled to a decreased bioavailability in arginine has also been documented in sputum, serum, erythrocytes, platelets and BALf samples from COPD patients and in animal models as well [59, 63]. This increase in arginase activity may contribute to pulmonary inflammation and airway remodelling [63] and low arginine levels seem to be associated with impaired smooth muscle relaxation [52]. In fact, arginine up-regulation is related with the body defense mechanism that increases endothelial NO production in lungs to reduce pulmonary hypertension associated with hypoxia [18]. Arginine depletion has also been correlated with pulmonary arterial vasoconstriction, occurring in response to venous thrombo embolism, leading to a decrease in NO production [76].

The lungs play another essential role in the regulation of arginine metabolism through the production and clearance of asymmetric dimethylarginine (ADMA), symmetric dimethylarginine (SDMA) and monomethylarginine (MMA) produced by arginine methyltransferase-mediated post-translational modifications [105]. These metabolites may compete with arginine potentially modulating its intracellular bioavailability. Moreover, ADMA and MMA act as endogenous competitive inhibitors of NOS [59]. Arginine availability to arginases and NOS under healthy homeostatic conditions is maintained through methylarginines clearance in the lung or kidney. However, higher levels of methylarginines in disease may reduce and/or redirect arginine metabolism [59]. A role for ADMA has been proposed in allergic asthma, obesity-induced asthma, obstructive sleep apnea and COPD. It could also be considered a marker of cardiovascular risk [52].

Glutamate/Glutamine Metabolism

Glutamine and glutamate, together with proline, histidine, arginine and ornithine, encompass 25% of the dietary amino acid intake and constitute the "glutamate family" of amino acids, which are disposed of through conversion to glutamate. In addition to its involvement in protein structure, glutamate plays critical roles in nutrition, metabolism and signaling [106]. According to its role as excitatory neurotransmitter, increased levels of plasma glutamate have been reported in patients affected by acute lung injury, such as ARDS, TB and high-altitude pulmonary edema (HAPE) [18, 69, 107]. Glutamate signaling can, in turn, induce NO production and also apoptosis through the production of caspase 3 [18]. A key role in mediating cytotoxicity is played by the glutamate agonist N-methyl-D-aspartate (NMDA) through NMDA receptors in the alveolar capillary area causing pulmonary edema [18].

Besides its role as the major fuel for rapidly replicating cells (*i.e.* immune cells and enterocytes), glutamine serves as a nontoxic carrier for ammonia. Being the most abundant amino acid in serum, it is largely present both in lungs and muscles [52]. The low levels of glutamine in muscles in acute disease states, such as injury and sepsis, suggest a decreased protein synthesis [108]. Conversely, increased glutamine concentration was observed in muscle and decreased in serum and urine of patients affected by chronic diseases (*i.e.* emphysema), in association with a decreased protein synthesis [27, 52, 108, 109]. These findings allow to speculate that both impaired outward glutamine transport and altered intracellular glutamine metabolism lead to a decrease of glutamine muscle efflux in COPD patients [109].

By modulating NO production, glutamine and glutamate can also condition the elastic response of the thoracic aortic wall [76]. In fact, endothelial formation of glutamate from the carbon atoms of leucine *via* the leucine-to-glutamate (Leu → Glu) signaling pathway increases cGMP to mediate relaxation [106]. Glutamine, in turn, inhibits the endothelial conversion of citrulline to arginine thus reducing the release of endothelium-derived relaxing factor and NO.

Glutamine is also used as a precursor of glucosamine-6-phosphate whose role as inhibitor of NO synthesis is well known [76]. The decrease of both amino acids observed in PAH patients is thought to be strictly related to the pulmonary arterial vasoconstriction occurring in response to venous thrombo embolism [76].

Other Amino Acids

Besides its role in proteinogenesis, the essential amino acid lysine is also involved in the crosslinking of collagen polypeptides, in the uptake of essential mineral nutrients and in the production of carnitine. Moreover, it is involved in histone modifications, thus affecting the epigenome. The finding of increased lysine levels in BALf from ARDS patients suggests a potential defensive role for this amino acid. It also triggers the decrease in NO levels thus increasing vascular resistance [18].

While the main role of glycine is to be the precursor of proteins, it can also play a key role as biosynthetic intermediate and neurotransmitter [110]. The finding of decreased serum levels of glycine in COPD patients with exacerbations compared to stable COPD subjects allowed to speculate that glycine concentration was inversely related to the degree of radiographic emphysema and cachexia [27]. However, as demonstrated in COPD subjects, in which it is positively correlated to inflammatory markers, *i.e.* C-reactive protein, glycine has also anti-inflammatory properties [27]. The decreased glycine levels reported in TB patients suggest for these individuals alterations in protein metabolism [64].

Glycine betaine (*N,N,N*-trimethylglycine), commonly referred to as betaine, and its precursor choline, are two compounds involved in the methylation of homocysteine to methionine and of DNA nitrogenous bases, processes in which betaine acts as a donor of methyl groups. Betaine is closely associated to protein and energy metabolism, to cellular nutrition and it promotes cell growth. It has been reported that plasma betaine level is correlated to cancer risk and it seems to be involved in the pathogens–host interaction during infection [83]. Betaine was suggested to have protective effect against endothelial dysfunction and atherosclerotic disease and this vascular protective activity was mainly attributed to homocysteine catabolism induction, oxidative stress reduction, and modulation of the inflammatory process involved in atherogenesis [76]. Decreased levels of both glycine betaine and choline have been observed in PAH and in TB patients [76, 83]. An increase of betaine and choline was reported in urine of COPD patients: since the conversion choline to betaine is also promoted by bacterial enzymes, this finding might be considered as a unique feature in the airways of these patients [52].

Proline, a proteogenic secondary amino acid is a key precursor of collagen formation and mesenchyme in the lungs. As a consequence of airway remodeling due to the activity of arginase from alveolar macrophages, a significant increase of proline was detected in BALf from subjects affected by ARDS [18]. The inflammatory state characterizing ARDS results in hypoxia, ROS production and oxidative stress. The exposure of proline residues in proteins to oxidation by ROS results in fragmentation and loss of protein function [18].

The gluconeogenic amino acid threonine was found to be associated with sepsis accounting to be a major aetiological factor of ARDS. In fact, it regulates immune responses through the production of antibodies and has a significant role in the inhibition of apoptosis by triggering lymphocyte proliferation [18].

While low plasma levels of the non-essential amino acid alanine (that seems to be associated with major adverse cardiovascular events) have been detected in COPD patients [98], the levels of same amino acid were increased in plasma of patients with SSc-related PAH [76]. Since alanine is a precursor of glutamate, increased levels of alanine associated to reduced glutamate concentrations in plasma appear to be consistent with a decreased alanine aminotransferase activity genetically determined [76].

Nitrogenous Bases, Nucleosides and Nucleotides

The presence of niacin (vitamin B3) metabolites in serum and urine from patients affected by COPD is well documented [52]. 1-methyl nicotinamide was shown to act as an anti-inflammatory molecule in asthma exacerbation by inducing the release of prostacyclin [52, 111]. Its levels were also further increased in patients affected by COPD [52]. Urinary concentrations of niacinamide and nicotinamide N-oxide were decreased in exacerbated COPD patients and, most likely, this reflects the reduced food intake (due to reduced appetite) that precedes exacerbations [27]. Trigonelline (1-methylnicotinic acid) is present in a number of plant beans, being produced during nicotinamide metabolism in case of oxidative stress. It is also endogenously formed through the methylation of the nitrogen atom of niacin [28]. Its levels were increased in relation to tobacco smoking and can be correlated with lung function in COPD patients [28, 52].

Uracil and pantothenate are involved in the coenzyme A biosynthesis pathway, which is critical in energy metabolism, and is relevant in the pathogenesis of asthma. Pantothenate levels increased by cigarette smoke have been observed in epithelial cell cultures [52].

Adenine is a nitrogenous base required for the production of adenosine, a purine nucleoside having a key role in cellular energy metabolism. Adenosine can be both pro-inflammatory and anti-inflammatory for mast cell stimulation. Increased levels of adenosine have been detected in patients with asthma in response to cellular damage [66].

The presence in urine of the metabolite hippurate is due to endogenous production by human cells and by the gut microbiome, as a result of the conjugation between benzoate and glycine. Urinary hippurate concentration is increased in response to inflammation and oxidative stress but it is also derived from the diet as the result of tea consumption [28]. COPD patients show increased levels of this metabolite in urine also in relation to their baseline lung function. This suggests that this correlation could be indicative of variable gut microbiome composition among individuals with differing lung function [28].

CONSIDERATIONS ON THE RELIABILITY AND ANALYSIS OF METABOLOMIC RESULTS IN THE STUDY OF RESPIRATORY DISORDERS

The careful reading of articles considered for the preparation of this chapter highlights unequivocally that the development of rigorous and standardized operating procedures for sample collection and pre-storage processing is an indispensable step to achieve reproducible results in respiratory metabolomics. The authors of these reports have underlined how the numerous factors that affect the metabolite profile of a biological fluid may represent obstacles to the correct interpretation of data. Storage temperature of samples is one of the factors that may influence metabolite composition. In most cases considered, samples have been stored at -80 °C mainly for two reasons. First of all, this temperature is expected to represent the best option for the preservation of a biofluid. Second, most laboratories are equipped with commercial instruments which operate at this temperature. However, not always the integrity of metabolites is completely preserved at this temperature [112]. It has been previously demonstrated that some analytes may undergo some loss (whose rate depends on the type of analyte/matrix) when left to a long-term storage [113]. This finding prompted the search of a strategy to assess the effect that different storage procedures have on the metabolite composition of the biofluid under investigation with the final purpose to preserve the "true" metabolic phenotype of the sample in the best possible way. The use of markers that are able to monitor sample integrity would certainly strengthen the reliability of data [114]. Despite the nature of respiratory diseases considered in articles commented in this chapter exhibited large deviations in pathological phenotype, results emerged from laboratory research could be linked to clinical aspects of the pathology. Metabolites that were over-/under-expressed as the presumed result of disorder well correlated across all cohorts of patients with their diseased state thereby providing a proof that data were reliably interpreted by investigators. This is an indirect confirmation of the suitability of NMR analysis as a tool that can offer a good clinical phenotyping of patients. Another point of interest addressed by authors concerns the strategies to account for both the: i) intra- and inter-individual variability and ii) the variability introduced by a number of potential confounding factors. These latter factors span from the diet to the smoking history and the age/gender composition of groups and are strongly problematic. Although matching of patients and controls could be a way for controlling/reducing variability, this approach often does not completely remove them. As a consequence, differences among patients could make it potentially difficult to understand whether the observed modifications in metabolite abundance, rather than representing the actual effect of the disease, are the result of unwanted confounding factors. Obviously, the key point to overcome this problem is to include in the study only patients characterized by the most uniform clinical/environmental conditions.

As previously anticipated, the NMR analytical platform has the advantage over other techniques to overcome the possible variability due to sample dilution. Data may be normalized by reference of the total metabolic content to the total spectra area of each sample [115].

Nevertheless, due to the wealth of signals present in spectra, deriving meaningful biological knowledge from the visual inspection of the entire dataset of metabolites produced by mono- and two-dimensional NMR experiments is indeed challenging. Interpretation of the acquired data still represents a significant bottleneck in a metabolomic experiment and statistical analyses appear the only procedures that enable extraction of a good amount of knowledge. Different statistical approaches (outlined above) have been applied to interpret the results described in the articles discussed in previous paragraphs. All methods have been identified by the authors as the most appropriate ones to analyze the numerous metabolomic data produced in their experiments. Not surprisingly, a few of these are common to all articles. This is the case of simple univariate statistical methods used for the identification of variables that are significant at a given probability *i.e.* the analysis of variance (ANOVA) and Student's t-test. However, a typical drawback connected with these methods is the presence of false-positives, particularly frequent within datasets containing hundreds/thousands of variables. To overcome these problems, multivariate statistical analyses have been extensively employed. Those most commonly used by the metabolomics community are unsupervised techniques such as principal component analysis (PCA) and supervised techniques such as partial least squares discriminant analysis (PLS-DA) and orthogonal partial least squares discriminant analysis (OPLS-DA) or, less frequently, bidirectional OPLS (O2PLS). To evaluate the effectiveness of these latter models, the R^2 (the percent of variation of the response predicted by the model according to the cross validation) and Q^2 (indicates how accurately the model can predict new data values) values have been determined. However, to improve the understanding of the mechanisms involved in the onset/progression of a respiratory disease, results of metabolomics experiments should enable identification of specific metabolic pathways that describe the biological perturbation under examination. Tools available to provide this identification, while being numerous, do not have the same power. The most powerful include the Kyoto Encyclopedia of Genes and Genomes (KEGG) [116, 117], the Ingenuity Pathway Analysis and the Human Metabolome Database [118, 119]. Interestingly, the KEGG database contains a useful new function, indicated as the "DISEASE" utility, whose role is to help researchers in the interpretation of metabolomics data generated from a range of diseases [120].

CONCLUDING REMARKS

The studies analyzed in this chapter have shown the power of NMR to classify and potentially diagnose patients suffering from respiratory conditions which span from disorders that are the major cause of global morbidity and mortality to diseases that are apparently less common/severe. The good number of metabolites detected enabled the identification of several metabolic pathways that describe the metabolic phenotype of respiratory diseases. We have observed that, while asthma can be characterized by the TCA cycle and histamine metabolism, COPD is characterized by BCAA catabolism and cystic fibrosis by glucose, purine and glutathione metabolism. Looking at the lists of metabolites identified, it is interesting to note that, while some of them are remodeled in multiple respiratory diseases compared to controls, the metabolic phenotype of each pathological condition is unique. The studies on COPD have also shown the ability of this procedure to demonstrate the influence of treatments on disorder progression.

The final purpose of metabolomics is to identify as many metabolites as possible to draw up an overview of disease-specific biochemistry. Despite NMR alone cannot provide the comprehensive coverage of all the metabolites contained in a biofluid, its application to respiratory diseases has indeed produced some promising results. Taken together, the data presented in this chapter confirm that metabolomics has the potential to diagnose respiratory diseases with a degree of accuracy higher than that of other methods.

Thus, it seems possible to state that the data so far acquired and those that will be collected with the future development of the techniques could enable researchers to greatly increase their ability to monitor a good number of lung diseases and, possibly, to predict patient response to interventions.

CONSENT FOR PUBLICATION

Not applicable.

ACKNOWLEDGEMENTS

The authors would like to thank Dr. Maddalena Cagnone, PhD in Biochemistry at the Department of Molecular Medicine, University of Pavia, for helpful discussion about the content of this chapter.

CONFLICT OF INTEREST

The authors declare no conflict of interest, financial or otherwise.

REFERENCES

[1]   Turi, K.N.; Romick-Rosendale, L.; Ryckman, K.K.; Hartert, T.V. A review of metabolomics approaches and their application in identifying causal pathways of childhood asthma. *J. Allergy Clin. Immunol.,* **2018**, *141*(4), 1191-1201.
[http://dx.doi.org/10.1016/j.jaci.2017.04.021] [PMID: 28479327]

[2]   Larive, C.K.; Larsen, S.C. NMR Developments and Applications. *Anal. Chem.,* **2017**, *89*(3), 1391.
[http://dx.doi.org/10.1021/acs.analchem.6b05082] [PMID: 28208241]

[3]   Lv, H.; Jiang, F.; Guan, D.; Lu, C.; Guo, B.; Chan, C.; Peng, S.; Liu, B.; Guo, W.; Zhu, H.; Xu, X.; Lu, A.; Zhang, G. Metabolomics and Its Application in the Development of Discovering Biomarkers for Osteoporosis Research. *Int. J. Mol. Sci.,* **2016**, *17*(12), 2018.
[http://dx.doi.org/10.3390/ijms17122018] [PMID: 27918446]

[4]   Psychogios, N.; Hau, D.D.; Peng, J.; Guo, A.C.; Mandal, R.; Bouatra, S.; Sinelnikov, I.; Krishnamurthy, R.; Eisner, R.; Gautam, B.; Young, N.; Xia, J.; Knox, C.; Dong, E.; Huang, P.; Hollander, Z.; Pedersen, T.L.; Smith, S.R.; Bamforth, F.; Greiner, R.; McManus, B.; Newman, J.W.; Goodfriend, T.; Wishart, D.S. The human serum metabolome. *PLoS One,* **2011**, *6*(2), e16957.
[http://dx.doi.org/10.1371/journal.pone.0016957] [PMID: 21359215]

[5]   Bouatra, S.; Aziat, F.; Mandal, R.; Guo, A.C.; Wilson, M.R.; Knox, C.; Bjorndahl, T.C.; Krishnamurthy, R.; Saleem, F.; Liu, P.; Dame, Z.T.; Poelzer, J.; Huynh, J.; Yallou, F.S.; Psychogios, N.; Dong, E.; Bogumil, R.; Roehring, C.; Wishart, D.S. The human urine metabolome. *PLoS One,* **2013**, *8*(9), e73076.
[http://dx.doi.org/10.1371/journal.pone.0073076] [PMID: 24023812]

[6]   Hao, L.; Wang, J.; Page, D.; Asthana, S.; Zetterberg, H.; Carlsson, C.; Okonkwo, O.C.; Li, L. Comparative Evaluation of MS-based Metabolomics Software and Its Application to Preclinical Alzheimer's Disease. *Sci. Rep.,* **2018**, *8*(1), 9291.
[http://dx.doi.org/10.1038/s41598-018-27031-x] [PMID: 29915347]

[7]   Yulianto, W.; Andarwulan, N.; Giriwono, P.E.; Pamungkas, J. HPLC-based metabolomics to identify cytotoxic compounds from Plectranthus amboinicus (Lour.) Spreng against human breast cancer MCF-7Cells. *J. Chromatogr. B Analyt. Technol. Biomed. Life Sci.,* **2016**, *1039*, 28-34.
[http://dx.doi.org/10.1016/j.jchromb.2016.10.024] [PMID: 27816313]

[8]   Huang, J.; Mo, J.; Zhao, G.; Lin, Q.; Wei, G.; Deng, W.; Chen, D.; Yu, B. Application of the amniotic fluid metabolome to the study of fetal malformations, using Down syndrome as a specific model. *Mol. Med. Rep.,* **2017**, *16*(5), 7405-7415.
[http://dx.doi.org/10.3892/mmr.2017.7507] [PMID: 28944830]

[9]   Tian, J.S.; Peng, G.J.; Wu, Y.F.; Zhou, J.J.; Xiang, H.; Gao, X.X.; Zhou, Y.Z.; Qin, X.M.; Du, G.H. A GC-MS urinary quantitative metabolomics analysis in depressed patients treated with TCM formula of Xiaoyaosan. *J. Chromatogr. B Analyt. Technol. Biomed. Life Sci.,* **2016**, *1026*, 227-235.
[http://dx.doi.org/10.1016/j.jchromb.2015.12.026] [PMID: 26733091]

[10]   Poisson, L.M.; Suhail, H.; Singh, J.; Datta, I.; Denic, A.; Labuzek, K.; Hoda, M.N.; Shankar, A.;

Kumar, A.; Cerghet, M.; Elias, S.; Mohney, R.P.; Rodriguez, M.; Rattan, R.; Mangalam, A.K.; Giri, S. Untargeted Plasma Metabolomics Identifies Endogenous Metabolite with Drug-like Properties in Chronic Animal Model of Multiple Sclerosis. *J. Biol. Chem.,* **2015**, *290*(52), 30697-30712.
[http://dx.doi.org/10.1074/jbc.M115.679068] [PMID: 26546682]

[11]　Xiong, X.; Sheng, X.; Liu, D.; Zeng, T.; Peng, Y.; Wang, Y. A GC/MS-based metabolomic approach for reliable diagnosis of phenylketonuria. *Anal. Bioanal. Chem.,* **2015**, *407*(29), 8825-8833.
[http://dx.doi.org/10.1007/s00216-015-9041-3] [PMID: 26410738]

[12]　Vignoli, A.; Ghini, V.; Meoni, G.; Licari, C.; Takis, P.G.; Tenori, L.; Turano, P.; Luchinat, C. High-throughput metabolomics by 1D NMR. *Angew. Chem. Int. Ed. Engl.,* **2018**. Epub ahead of print.
[PMID: 29999221]

[13]　Irwin, C.; van Reenen, M.; Mason, S.; Mienie, L.J.; Wevers, R.A.; Westerhuis, J.A.; Reinecke, C.J. The 1H-NMR-based metabolite profile of acute alcohol consumption: A metabolomics intervention study. *PLoS One,* **2018**, *13*(5), e0196850.
[http://dx.doi.org/10.1371/journal.pone.0196850] [PMID: 29746531]

[14]　Veenstra, T.D. Metabolomics: the final frontier? *Genome Med.,* **2012**, *4*(4), 40.
[http://dx.doi.org/10.1186/gm339] [PMID: 22546050]

[15]　Ramsden, J. *J. Bioinformatics, Metabolomics and Metabonomics*; Springer-Verlag: London, **2009**, pp. 239-250.

[16]　Ciaramelli, C.; Fumagalli, M.; Viglio, S.; Bardoni, A.M.; Piloni, D.; Meloni, F.; Iadarola, P.; Airoldi, C. ^{1}H NMR to evaluate the metabolome of bronchoalveolar lavage fluid (BALf) in bronchiolitis obliterans syndrome (BOS): Toward the development of a new approach for biomarker identification. *J. Proteome Res.,* **2017**, *16*(4), 1669-1682.
[http://dx.doi.org/10.1021/acs.jproteome.6b01038] [PMID: 28245130]

[17]　Wolak, J.E.; Esther, C.R., Jr; O'Connell, T.M. Metabolomic analysis of bronchoalveolar lavage fluid from cystic fibrosis patients. *Biomarkers,* **2009**, *14*(1), 55-60.
[http://dx.doi.org/10.1080/13547500802688194] [PMID: 19283525]

[18]　Viswan, A.; Singh, C.; Rai, R.K.; Azim, A.; Sinha, N.; Baronia, A.K. Metabolomics based predictive biomarker model of ARDS. A systemic measure of clinical hypoxemia. *PLoS One,* **2017**, *12*(11), e0187545.
[http://dx.doi.org/10.1371/journal.pone.0187545] [PMID: 29095932]

[19]　Viswan, A.; Sharma, R.K.; Azim, A.; Sinha, N. NMR-Based Metabolic Snapshot from Minibronchoalveolar Lavage Fluid: An Approach To Unfold Human Respiratory Metabolomics. *J. Proteome Res.,* **2016**, *15*(1), 302-310.
[http://dx.doi.org/10.1021/acs.jproteome.5b00919] [PMID: 26587756]

[20]　Iadarola, P.; Viglio, S. Spit it out! How could the sputum proteome aid clinical research into pulmonary diseases? *Expert Rev. Proteomics,* **2017**, *14*(5), 391-393.
[http://dx.doi.org/10.1080/14789450.2017.1317246] [PMID: 28388247]

[21]　Terracciano, R.; Pelaia, G.; Preianò, M.; Savino, R. Asthma and COPD proteomics: current approaches and future directions. *Proteomics Clin. Appl.,* **2015**, *9*(1-2), 203-220.
[http://dx.doi.org/10.1002/prca.201400099] [PMID: 25504544]

[22]　Duchemann, B.; Triba, M.N.; Guez, D.; Rzeznik, M.; Savarin, P.; Nunes, H.; Valeyre, D.; Bernaudin, J.F.; Le Moyec, L. Nuclear magnetic resonance spectroscopic analysis of salivary metabolome in sarcoidosis. *Sarcoidosis Vasc. Diffuse Lung Dis.,* **2016**, *33*(1), 10-16.
[PMID: 27055831]

[23]　Romano, F.; Meoni, G.; Manavella, V.; Baima, G.; Tenori, L.; Cacciatore, S.; Aimetti, M. Analysis of salivary phenotypes of generalized aggressive and chronic periodontitis through nuclear magnetic resonance-based metabolomics. *J. Periodontol.,* **2018**, *89*(12), 1452-1460. Epub ahead of print.
[http://dx.doi.org/10.1002/JPER.18-0097] [PMID: 29877582]

[24]　Nagana Gowda, G.A.; Raftery, D. Quantitating metabolites in protein precipitated serum using NMR spectroscopy. *Anal. Chem.,* **2014**, *86*(11), 5433-5440.
[http://dx.doi.org/10.1021/ac5005103] [PMID: 24796490]

[25]　Forsythe, I.J.; Wishart, D.S. Exploring human metabolites using the human metabolome database. *Curr. Protoc. Bioinformatics,* **2009**, *14*, 8.1.
[http://dx.doi.org/10.1002/0471250953.bi1408s25]

[26]　Nobakht M Gh, B.F.; Aliannejad, R.; Rezaei-Tavirani, M.; Taheri, S.; Oskouie, A.A. The metabolomics of airway diseases, including COPD, asthma and cystic fibrosis. *Biomarkers,* **2015**, *20*(1), 5-16.
[http://dx.doi.org/10.3109/1354750X.2014.983167] [PMID: 25403491]

[27]　Fortis, S.; Lusczek, E.R.; Weinert, C.R.; Beilman, G.J. Metabolomics in COPD Acute Respiratory Failure Requiring Noninvasive Positive Pressure Ventilation. *Can. Respir. J.,* **2017**, *2017*, 9480346.
[http://dx.doi.org/10.1155/2017/9480346] [PMID: 29391845]

[28]　McClay, J.L.; Adkins, D.E.; Isern, N.G.; O'Connell, T.M.; Wooten, J.B.; Zedler, B.K.; Dasika, M.S.; Webb, B.T.; Webb-Robertson, B.J.; Pounds, J.G.; Murrelle, E.L.; Leppert, M.F.; van den Oord, E.J. (1)H nuclear magnetic resonance metabolomics analysis identifies novel urinary biomarkers for lung function. *J. Proteome Res.,* **2010**, *9*(6), 3083-3090.
[http://dx.doi.org/10.1021/pr1000048] [PMID: 20408573]

[29]　Kan, M.; Shumyatcher, M.; Himes, B.E. Using omics approaches to understand pulmonary diseases. *Respir. Res.,* **2017**, *18*(1), 149.
[http://dx.doi.org/10.1186/s12931-017-0631-9] [PMID: 28774304]

[30]　Wang, L.; Tang, Y.; Liu, S.; Mao, S.; Ling, Y.; Liu, D.; He, X.; Wang, X. Metabonomic profiling of serum and urine by (1)H NMR-based spectroscopy discriminates patients with chronic obstructive pulmonary disease and healthy individuals. *PLoS One,* **2013**, *8*(6), e65675.
[http://dx.doi.org/10.1371/journal.pone.0065675] [PMID: 23755267]

[31]　Quan-Jun, Y.; Jian-Ping, Z.; Jian-Hua, Z.; Yong-Long, H.; Bo, X.; Jing-Xian, Z.; Bona, D.; Yuan, Z.; Cheng, G. Distinct Metabolic Profile of Inhaled Budesonide and Salbutamol in Asthmatic Children during Acute Exacerbation. *Basic Clin. Pharmacol. Toxicol.,* **2017**, *120*(3), 303-311.
[http://dx.doi.org/10.1111/bcpt.12686] [PMID: 27730746]

[32]　Ząbek, A.; Stanimirova, I.; Deja, S.; Barg, W.; Kowal, A.; Korzeniewska, A.; Orczyk-Pawiłowicz, M.; Baranowski, D.; Gdaniec, Z.; Jankowska, R.; Młynarz, P. Fusion of the ¹H NMR data of serum, urine and exhaled breath condensate in order to discriminate chronic obstructive pulmonary disease and obstructive sleep apnea syndrome. *Metabolomics,* **2015**, *11*(6), 1563-1574.
[http://dx.doi.org/10.1007/s11306-015-0808-5] [PMID: 26491417]

[33]　Beckonert, O.; Keun, H.C.; Ebbels, T.M.; Bundy, J.; Holmes, E.; Lindon, J.C.; Nicholson, J.K. Metabolic profiling, metabolomic and metabonomic procedures for NMR spectroscopy of urine, plasma, serum and tissue extracts. *Nat. Protoc.,* **2007**, *2*(11), 2692-2703.
[http://dx.doi.org/10.1038/nprot.2007.376] [PMID: 18007604]

[34]　Snowden, S.; Dahlén, S.E.; Wheelock, C.E. Application of metabolomics approaches to the study of respiratory diseases. *Bioanalysis,* **2012**, *4*(18), 2265-2290.
[http://dx.doi.org/10.4155/bio.12.218] [PMID: 23046268]

[35]　Friebolin, H. *Basic One- and Two-Dimensional NMR Spectroscopy,* 5th ed; John Wiley & Sons Inc.: New York, **2010**.

[36]　Keeler, J. *Understanding NMR Spectroscopy,* 2nd ed; John Wiley & Sons Inc.: New York, **2010**.

[37]　Li, M.; Riddle, S.; Zhang, H.; D'Alessandro, A.; Flockton, A.; Serkova, N.J.; Hansen, K.C.; Moldvan, R.; McKeon, B.A.; Frid, M.; Kumar, S.; Li, H.; Liu, H.; Caánovas, A.; Medrano, J.F.; Thomas, M.G.; Iloska, D.; Plecitá-Hlavatá, L.; Ježek, P.; Pullamsetti, S.; Fini, M.A.; El Kasmi, K.C.; Zhang, Q.; Stenmark, K.R. Metabolic Reprogramming Regulates the Proliferative and Inflammatory Phenotype of

Adventitial Fibroblasts in Pulmonary Hypertension Through the Transcriptional Corepressor C-Terminal Binding Protein-1. *Circulation*, **2016**, *134*(15), 1105-1121.
[http://dx.doi.org/10.1161/CIRCULATIONAHA.116.023171] [PMID: 27562971]

[38] Tzika, A.A.; Constantinou, C.; Bandyopadhaya, A.; Psychogios, N.; Lee, S.; Mindrinos, M.; Martyn, J.A.; Tompkins, R.G.; Rahme, L.G. A small volatile bacterial molecule triggers mitochondrial dysfunction in murine skeletal muscle. *PLoS One*, **2013**, *8*(9), e74528.
[http://dx.doi.org/10.1371/journal.pone.0074528] [PMID: 24098655]

[39] Markley, J.L.; Brüschweiler, R.; Edison, A.S.; Eghbalnia, H.R.; Powers, R.; Raftery, D.; Wishart, D.S. The future of NMR-based metabolomics. *Curr. Opin. Biotechnol.,* **2017**, *43*, 34-40.
[http://dx.doi.org/10.1016/j.copbio.2016.08.001] [PMID: 27580257]

[40] Motta, A.; Paris, D.; D'Amato, M.; Melck, D.; Calabrese, C.; Vitale, C.; Stanziola, A.A.; Corso, G.; Sofia, M.; Maniscalco, M. NMR metabolomic analysis of exhaled breath condensate of asthmatic patients at two different temperatures. *J. Proteome Res.,* **2014**, *13*(12), 6107-6120.
[http://dx.doi.org/10.1021/pr5010407] [PMID: 25393672]

[41] Jacobsen, N.E. *NMR Spectroscopy Explained: Simplified Theory, Applications and Examples for Organic Chemistry and Structural Biology*; John Wiley & Sons Inc: New York, **2007**.
[http://dx.doi.org/10.1002/9780470173350]

[42] Simpson, J.H. *Organic Structure Determination Using 2-D NMR Spectroscopy. A Problem-Based Approach,* 2nd ed; Academic Press: New York, **2012**.

[43] de Laurentiis, G.; Paris, D.; Melck, D.; Montuschi, P.; Maniscalco, M.; Bianco, A.; Sofia, M.; Motta, A. Separating smoking-related diseases using NMR-based metabolomics of exhaled breath condensate. *J. Proteome Res.,* **2013**, *12*(3), 1502-1511.
[http://dx.doi.org/10.1021/pr301171p] [PMID: 23360153]

[44] Atzei, A.; Atzori, L.; Moretti, C.; Barberini, L.; Noto, A.; Ottonello, G.; Pusceddu, E.; Fanos, V. Metabolomics in paediatric respiratory diseases and bronchiolitis. *J. Matern. Fetal Neonatal Med.,* **2011**, *24*(2) Suppl. 2, 59-62.
[http://dx.doi.org/10.3109/14767058.2011.607012] [PMID: 21966897]

[45] Ripps, H.; Shen, W. Review: taurine: a "very essential" amino acid. *Mol. Vis.,* **2012**, *18*, 2673-2686.
[PMID: 23170060]

[46] Lambert, I.H.; Kristensen, D.M.; Holm, J.B.; Mortensen, O.H. Physiological role of taurine--from organism to organelle. *Acta Physiol. (Oxf.),* **2015**, *213*(1), 191-212.
[http://dx.doi.org/10.1111/apha.12365] [PMID: 25142161]

[47] Lang, P.A.; Warskulat, U.; Heller-Stilb, B.; Huang, D.Y.; Grenz, A.; Myssina, S.; Duszenko, M.; Lang, F.; Häussinger, D.; Vallon, V.; Wieder, T. Blunted apoptosis of erythrocytes from taurine transporter deficient mice. *Cell. Physiol. Biochem.,* **2003**, *13*(6), 337-346.
[http://dx.doi.org/10.1159/000075121] [PMID: 14631140]

[48] Schuller-Levis, G.B.; Park, E. Taurine: new implications for an old amino acid. *FEMS Microbiol. Lett.,* **2003**, *226*(2), 195-202.
[http://dx.doi.org/10.1016/S0378-1097(03)00611-6] [PMID: 14553911]

[49] Lambert, I.H.; Hoffmann, E.K.; Pedersen, S.F. Cell volume regulation: physiology and pathophysiology. *Acta Physiol. (Oxf.),* **2008**, *194*(4), 255-282.
[http://dx.doi.org/10.1111/j.1748-1716.2008.01910.x] [PMID: 18945273]

[50] Gallos, G.; Yim, P.; Chang, S.; Zhang, Y.; Xu, D.; Cook, J.M.; Gerthoffer, W.T.; Emala, C.W., Sr Targeting the restricted α-subunit repertoire of airway smooth muscle GABAA receptors augments airway smooth muscle relaxation. *Am. J. Physiol. Lung Cell. Mol. Physiol.,* **2012**, *302*(2), L248-L256.
[http://dx.doi.org/10.1152/ajplung.00131.2011] [PMID: 21949156]

[51] Lambert, I.H. Regulation of the cellular content of the organic osmolyte taurine in mammalian cells. *Neurochem. Res.,* **2004**, *29*(1), 27-63.

[http://dx.doi.org/10.1023/B:NERE.0000010433.08577.96] [PMID: 14992263]

[52] Adamko, D.J.; Nair, P.; Mayers, I.; Tsuyuki, R.T.; Regush, S.; Rowe, B.H. Metabolomic profiling of asthma and chronic obstructive pulmonary disease: A pilot study differentiating diseases. *J. Allergy Clin. Immunol.,* **2015**, *136*(3), 571-580.e3.
[http://dx.doi.org/10.1016/j.jaci.2015.05.022] [PMID: 26152317]

[53] Wilkinson, T.M.; Patel, I.S.; Wilks, M.; Donaldson, G.C.; Wedzicha, J.A. Airway bacterial load and FEV1 decline in patients with chronic obstructive pulmonary disease. *Am. J. Respir. Crit. Care Med.,* **2003**, *167*(8), 1090-1095.
[http://dx.doi.org/10.1164/rccm.200210-1179OC] [PMID: 12684248]

[54] Oppermann, F.B.; Steinbüchel, A. Identification and molecular characterization of the aco genes encoding the Pelobacter carbinolicus acetoin dehydrogenase enzyme system. *J. Bacteriol.,* **1994**, *176*(2), 469-485.
[http://dx.doi.org/10.1128/jb.176.2.469-485.1994] [PMID: 8110297]

[55] Xiao, Z.; Xu, P. Acetoin metabolism in bacteria. *Crit. Rev. Microbiol.,* **2007**, *33*(2), 127-140.
[http://dx.doi.org/10.1080/10408410701364604] [PMID: 17558661]

[56] Airoldi, C.; Ciaramelli, C.; Fumagalli, M.; Bussei, R.; Mazzoni, V.; Viglio, S.; Iadarola, P.; Stolk, J. (1)H NMR to explore the metabolome of exhaled breath condensate in α(1)-antitrypsin deficient patients: A Pilot Study. *J. Proteome Res.,* **2016**, *15*(12), 4569-4578.
[http://dx.doi.org/10.1021/acs.jproteome.6b00648] [PMID: 27646345]

[57] Montuschi, P.; Santini, G.; Mores, N.; Vignoli, A.; Macagno, F.; Shoreh, R.; Tenori, L.; Zini, G.; Fuso, L.; Mondino, C.; Di Natale, C.; D'Amico, A.; Luchinat, C.; Barnes, P.J.; Higenbottam, T. Breathomics for assessing the effects of treatment and withdrawal with inhaled beclomethasone/formoterol in patients with COPD. *Front. Pharmacol.,* **2018**, *9*, 258.
[http://dx.doi.org/10.3389/fphar.2018.00258] [PMID: 29719507]

[58] Luster, A.D.; Alon, R.; von Andrian, U.H. Immune cell migration in inflammation: present and future therapeutic targets. *Nat. Immunol.,* **2005**, *6*(12), 1182-1190.
[http://dx.doi.org/10.1038/ni1275] [PMID: 16369557]

[59] Lara, A.; Khatri, S.B.; Wang, Z.; Comhair, S.A.; Xu, W.; Dweik, R.A.; Bodine, M.; Levison, B.S.; Hammel, J.; Bleecker, E.; Busse, W.; Calhoun, W.J.; Castro, M.; Chung, K.F.; Curran-Everett, D.; Gaston, B.; Israel, E.; Jarjour, N.; Moore, W.; Peters, S.P.; Teague, W.G.; Wenzel, S.; Hazen, S.L.; Erzurum, S.C. Alterations of the arginine metabolome in asthma. *Am. J. Respir. Crit. Care Med.,* **2008**, *178*(7), 673-681.
[http://dx.doi.org/10.1164/rccm.200710-1542OC] [PMID: 18635886]

[60] Boisvert, F.M.; Richard, S. Arginine methylation regulates the cytokine response. *Mol. Cell,* **2004**, *15*(4), 492-494.
[http://dx.doi.org/10.1016/j.molcel.2004.08.011] [PMID: 15327764]

[61] Schwartz, D.A. Epigenetics and environmental lung disease. *Proc. Am. Thorac. Soc.,* **2010**, *7*(2), 123-125.
[http://dx.doi.org/10.1513/pats.200908-084RM] [PMID: 20427583]

[62] Jung, J.; Kim, S.H.; Lee, H.S.; Choi, G.S.; Jung, Y.S.; Ryu, D.H.; Park, H.S.; Hwang, G.S. Serum metabolomics reveals pathways and biomarkers associated with asthma pathogenesis. *Clin. Exp. Allergy,* **2013**, *43*(4), 425-433.
[http://dx.doi.org/10.1111/cea.12089] [PMID: 23517038]

[63] Singh, B.; Jana, S.K.; Ghosh, N.; Das, S.K.; Joshi, M.; Bhattacharyya, P.; Chaudhury, K. Metabolomic profiling of doxycycline treatment in chronic obstructive pulmonary disease. *J. Pharm. Biomed. Anal.,* **2017**, *132*, 103-108.
[http://dx.doi.org/10.1016/j.jpba.2016.09.034] [PMID: 27697570]

[64] Zhou, A.; Ni, J.; Xu, Z.; Wang, Y.; Lu, S.; Sha, W.; Karakousis, P.C.; Yao, Y.F. Application of (1)h NMR spectroscopy-based metabolomics to sera of tuberculosis patients. *J. Proteome Res.,* **2013**,

12(10), 4642-4649.
[http://dx.doi.org/10.1021/pr4007359] [PMID: 23980697]

[65]   Koeslag, J.H.; Noakes, T.D.; Sloan, A.W. Post-exercise ketosis. *J. Physiol.,* **1980**, *301*, 79-90.
[http://dx.doi.org/10.1113/jphysiol.1980.sp013190] [PMID: 6997456]

[66]   Saude, E.J.; Skappak, C.D.; Regush, S.; Cook, K.; Ben-Zvi, A.; Becker, A.; Moqbel, R.; Sykes, B.D.;
Rowe, B.H.; Adamko, D.J. Metabolomic profiling of asthma: diagnostic utility of urine nuclear
magnetic resonance spectroscopy. *J. Allergy Clin. Immunol.,* **2011**, *127*(3), 757-64.e1, 6.
[http://dx.doi.org/10.1016/j.jaci.2010.12.1077] [PMID: 21377043]

[67]   Wang, C.; Li, J.X.; Tang, D.; Zhang, J.Q.; Fang, L.Z.; Fu, W.P.; Liu, L.; Dai, L.M. Metabolic changes
of different high-resolution computed tomography phenotypes of COPD after budesonide-formoterol
treatment. *Int. J. Chron. Obstruct. Pulmon. Dis.,* **2017**, *12*, 3511-3521.
[http://dx.doi.org/10.2147/COPD.S152134] [PMID: 29255358]

[68]   Ubhi, B.K.; Cheng, K.K.; Dong, J.; Janowitz, T.; Jodrell, D.; Tal-Singer, R.; MacNee, W.; Lomas,
D.A.; Riley, J.H.; Griffin, J.L.; Connor, S.C. Targeted metabolomics identifies perturbations in amino
acid metabolism that sub-classify patients with COPD. *Mol. Biosyst.,* **2012**, *8*(12), 3125-3133.
[http://dx.doi.org/10.1039/c2mb25194a] [PMID: 23051772]

[69]   Zhou, A.; Ni, J.; Xu, Z.; Wang, Y.; Zhang, H.; Wu, W.; Lu, S.; Karakousis, P.C.; Yao, Y.F.
Metabolomics specificity of tuberculosis plasma revealed by (1)H NMR spectroscopy. *Tuberculosis
(Edinb.),* **2015**, *95*(3), 294-302.
[http://dx.doi.org/10.1016/j.tube.2015.02.038] [PMID: 25736521]

[70]   Tom, A.; Nair, K.S. Assessment of branched-chain amino Acid status and potential for biomarkers. *J.
Nutr.,* **2006**, *136*(1) Suppl., 324S-330S.
[http://dx.doi.org/10.1093/jn/136.1.324S] [PMID: 16365107]

[71]   Puente-Maestu, L.; Pérez-Parra, J.; Godoy, R.; Moreno, N.; Tejedor, A.; González-Aragoneses, F.;
Bravo, J.L.; Alvarez, F.V.; Camaño, S.; Agustí, A. Abnormal mitochondrial function in locomotor and
respiratory muscles of COPD patients. *Eur. Respir. J.,* **2009**, *33*(5), 1045-1052.
[http://dx.doi.org/10.1183/09031936.00112408] [PMID: 19129279]

[72]   Bertini, I.; Luchinat, C.; Miniati, M.; Monti, S.; Tenori, L. Phenotyping COPD by 1H NMR
metabolomics of exhaled breath condensate. *Metabolomics,* **2014**, *10*, 302-311.
[http://dx.doi.org/10.1007/s11306-013-0572-3]

[73]   Deja, S.; Porebska, I.; Kowal, A.; Zabek, A.; Barg, W.; Pawelczyk, K.; Stanimirova, I.; Daszykowski,
M.; Korzeniewska, A.; Jankowska, R.; Mlynarz, P. Metabolomics provide new insights on lung cancer
staging and discrimination from chronic obstructive pulmonary disease. *J. Pharm. Biomed. Anal.,*
2014, *100*, 369-380.
[http://dx.doi.org/10.1016/j.jpba.2014.08.020] [PMID: 25213261]

[74]   Tan, L.C.; Yang, W.J.; Fu, W.P.; Su, P.; Shu, J.K.; Dai, L.M. ^{1}H-NMR-based metabolic profiling of
healthy individuals and high-resolution CT-classified phenotypes of COPD with treatment of
tiotropium bromide. *Int. J. Chron. Obstruct. Pulmon. Dis.,* **2018**, *13*, 2985-2997.
[http://dx.doi.org/10.2147/COPD.S173264] [PMID: 30310274]

[75]   Montuschi, P.; Paris, D.; Melck, D.; Lucidi, V.; Ciabattoni, G.; Raia, V.; Calabrese, C.; Bush, A.;
Barnes, P.J.; Motta, A. NMR spectroscopy metabolomic profiling of exhaled breath condensate in
patients with stable and unstable cystic fibrosis. *Thorax,* **2012**, *67*(3), 222-228.
[http://dx.doi.org/10.1136/thoraxjnl-2011-200072] [PMID: 22106016]

[76]   Deidda, M.; Piras, C.; Cadeddu Dessalvi, C.; Locci, E.; Barberini, L.; Orofino, S.; Musu, M.; Mura,
M.N.; Manconi, P.E.; Finco, G.; Atzori, L.; Mercuro, G. Distinctive metabolomic fingerprint in
scleroderma patients with pulmonary arterial hypertension. *Int. J. Cardiol.,* **2017**, *241*, 401-406.
[http://dx.doi.org/10.1016/j.ijcard.2017.04.024] [PMID: 28476520]

[77]   Jensen, P.R.; Peitersen, T.; Karlsson, M.; In 't Zandt, R.; Gisselsson, A.; Hansson, G.; Meier, S.;
Lerche, M.H. Tissue-specific short chain fatty acid metabolism and slow metabolic recovery after

ischemia from hyperpolarized NMR in vivo. *J. Biol. Chem.,* **2009**, *284*(52), 36077-36082.
[http://dx.doi.org/10.1074/jbc.M109.066407] [PMID: 19861411]

[78]   Hellerstein, M.K.; Benowitz, N.L.; Neese, R.A.; Schwartz, J.M.; Hoh, R.; Jacob, P., III; Hsieh, J.;
Faix, D. Effects of cigarette smoking and its cessation on lipid metabolism and energy expenditure in
heavy smokers. *J. Clin. Invest.,* **1994**, *93*(1), 265-272.
[http://dx.doi.org/10.1172/JCI116955] [PMID: 8282797]

[79]   Geamanu, A.; Gupta, S.V.; Bauerfeld, C.; Samavati, L. Metabolomics connects aberrant bioenergetic,
transmethylation, and gut microbiota in sarcoidosis. *Metabolomics,* **2016**, *12*, 35.
[http://dx.doi.org/10.1007/s11306-015-0932-2] [PMID: 27489531]

[80]   Tannahill, G.M.; Curtis, A.M.; Adamik, J.; Palsson-McDermott, E.M.; McGettrick, A.F.; Goel, G.;
Frezza, C.; Bernard, N.J.; Kelly, B.; Foley, N.H.; Zheng, L.; Gardet, A.; Tong, Z.; Jany, S.S.; Corr,
S.C.; Haneklaus, M.; Caffrey, B.E.; Pierce, K.; Walmsley, S.; Beasley, F.C.; Cummins, E.; Nizet, V.;
Whyte, M.; Taylor, C.T.; Lin, H.; Masters, S.L.; Gottlieb, E.; Kelly, V.P.; Clish, C.; Auron, P.E.;
Xavier, R.J.; O'Neill, L.A. Succinate is an inflammatory signal that induces IL-1β through HIF-1α.
Nature, **2013**, *496*(7444), 238-242.
[http://dx.doi.org/10.1038/nature11986] [PMID: 23535595]

[81]   Lower, E.E.; Malhotra, A.; Sudurlescu, V.; Baughman, R.P. Sarcoidosis, fatigue, and sleep apnea.
Chest, **2013**, *144*(6), 1976-1977.
[http://dx.doi.org/10.1378/chest.13-1761] [PMID: 24297143]

[82]   O'Neill, L.A.; Hardie, D.G. Metabolism of inflammation limited by AMPK and pseudo-starvation.
Nature, **2013**, *493*(7432), 346-355.
[http://dx.doi.org/10.1038/nature11862] [PMID: 23325217]

[83]   Sun, L.; Li, J.Q.; Ren, N.; Qi, H.; Dong, F.; Xiao, J.; Xu, F.; Jiao, W.W.; Shen, C.; Song, W.Q.; Shen,
A.D. Utility of Novel Plasma Metabolic Markers in the Diagnosis of Pediatric Tuberculosis: A
Classification and Regression Tree Analysis Approach. *J. Proteome Res.,* **2016**, *15*(9), 3118-3125.
[http://dx.doi.org/10.1021/acs.jproteome.6b00228] [PMID: 27451809]

[84]   Blumenthal, A.; Isovski, F.; Rhee, K.Y. Tuberculosis and host metabolism: ancient associations, fresh
insights. *Transl. Res.,* **2009**, *154*(1), 7-14.
[http://dx.doi.org/10.1016/j.trsl.2009.04.004] [PMID: 19524868]

[85]   Zhivotovsky, B.; Orrenius, S. The Warburg Effect returns to the cancer stage. *Semin. Cancer Biol.,*
2009, *19*(1), 1-3.
[http://dx.doi.org/10.1016/j.semcancer.2008.12.003] [PMID: 19162190]

[86]   Cheng, L.L.; Chang, I.W.; Smith, B.L.; Gonzalez, R.G. Evaluating human breast ductal carcinomas
with high-resolution magic-angle spinning proton magnetic resonance spectroscopy. *J. Magn. Reson.,*
1998, *135*(1), 194-202.
[http://dx.doi.org/10.1006/jmre.1998.1578] [PMID: 9799694]

[87]   Muñoz-Elías, E.J.; Upton, A.M.; Cherian, J.; McKinney, J.D. Role of the methylcitrate cycle in
Mycobacterium tuberculosis metabolism, intracellular growth, and virulence. *Mol. Microbiol.,* **2006**,
60(5), 1109-1122.
[http://dx.doi.org/10.1111/j.1365-2958.2006.05155.x] [PMID: 16689789]

[88]   Chang, C.H.; Curtis, J.D.; Maggi, L.B., Jr; Faubert, B.; Villarino, A.V.; O'Sullivan, D.; Huang, S.C.;
van der Windt, G.J.; Blagih, J.; Qiu, J.; Weber, J.D.; Pearce, E.J.; Jones, R.G.; Pearce, E.L.
Posttranscriptional control of T cell effector function by aerobic glycolysis. *Cell,* **2013**, *153*(6), 1239-
1251.
[http://dx.doi.org/10.1016/j.cell.2013.05.016] [PMID: 23746840]

[89]   Adeva-Andany, M.; López-Ojén, M.; Funcasta-Calderón, R.; Ameneiros-Rodríguez, E.; Donapetry-
García, C.; Vila-Altesor, M.; Rodríguez-Seijas, J. Comprehensive review on lactate metabolism in
human health. *Mitochondrion,* **2014**, *17*, 76-100.
[http://dx.doi.org/10.1016/j.mito.2014.05.007] [PMID: 24929216]

[90]   Meert, K.L.; McCaulley, L.; Sarnaik, A.P. Mechanism of lactic acidosis in children with acute severe asthma. *Pediatr. Crit. Care Med.,* **2012**, *13*(1), 28-31.
[http://dx.doi.org/10.1097/PCC.0b013e3182196aa2] [PMID: 21460758]

[91]   Pechlivanis, A.; Kostidis, S.; Saraslanidis, P.; Petridou, A.; Tsalis, G.; Mougios, V.; Gika, H.G.; Mikros, E.; Theodoridis, G.A. (1)H NMR-based metabonomic investigation of the effect of two different exercise sessions on the metabolic fingerprint of human urine. *J. Proteome Res.,* **2010**, *9*(12), 6405-6416.
[http://dx.doi.org/10.1021/pr100684t] [PMID: 20932058]

[92]   Nicholson, J.K.; Timbrell, J.A.; Sadler, P.J. Proton NMR spectra of urine as indicators of renal damage. Mercury-induced nephrotoxicity in rats. *Mol. Pharmacol.,* **1985**, *27*(6), 644-651.
[PMID: 2860559]

[93]   Bairaktari, E.; Seferiadis, K.; Liamis, G.; Psihogios, N.; Tsolas, O.; Elisaf, M. Rhabdomyolysis-related renal tubular damage studied by proton nuclear magnetic resonance spectroscopy of urine. *Clin. Chem.,* **2002**, *48*(7), 1106-1109.
[PMID: 12089184]

[94]   Sue, D.Y.; Wasserman, K.; Moricca, R.B.; Casaburi, R. Metabolic acidosis during exercise in patients with chronic obstructive pulmonary disease. Use of the V-slope method for anaerobic threshold determination. *Chest,* **1988**, *94*(5), 931-938.
[http://dx.doi.org/10.1378/chest.94.5.931] [PMID: 3180897]

[95]   Worlitzsch, D.; Tarran, R.; Ulrich, M.; Schwab, U.; Cekici, A.; Meyer, K.C.; Birrer, P.; Bellon, G.; Berger, J.; Weiss, T.; Botzenhart, K.; Yankaskas, J.R.; Randell, S.; Boucher, R.C.; Döring, G. Effects of reduced mucus oxygen concentration in airway Pseudomonas infections of cystic fibrosis patients. *J. Clin. Invest.,* **2002**, *109*(3), 317-325.
[http://dx.doi.org/10.1172/JCI0213870] [PMID: 11827991]

[96]   Monirujjaman, M.; Ferdouse, A. Metabolic and Physiological Roles of Branched-Chain Amino Acids. *Adv. Mol. Biol.,* **2014**, *2014*, 1-6.
[http://dx.doi.org/10.1155/2014/364976]

[97]   Babchia, N.; Calipel, A.; Mouriaux, F.; Faussat, A.M.; Mascarelli, F. The PI3K/Akt and mTOR/P70S6K signaling pathways in human uveal melanoma cells: interaction with B-Raf/ERK. *Invest. Ophthalmol. Vis. Sci.,* **2010**, *51*(1), 421-429.
[http://dx.doi.org/10.1167/iovs.09-3974] [PMID: 19661225]

[98]   Rodrıguez, D.A.; Alcarraz-Vizan, G.; Dıaz-Moralli, S.; Reed, M.; Gomez, F.P.; Falciani, F.; Günther, U.; Roca, J.; Cascante, M. Plasma metabolic profile in COPD patients: effects of exerciseand endurance training. *Metabolomics,* **2012**, *8*, 508-516.
[http://dx.doi.org/10.1007/s11306-011-0336-x]

[99]   Piscopo, P.; Crestini, A.; Adduci, A.; Ferrante, A.; Massari, M.; Popoli, P.; Vanacore, N.; Confaloni, A. Altered oxidative stress profile in the cortex of mice fed an enriched branched-chain amino acids diet: possible link with amyotrophic lateral sclerosis? *J. Neurosci. Res.,* **2011**, *89*(8), 1276-1283.
[http://dx.doi.org/10.1002/jnr.22655] [PMID: 21538464]

[100] Ohtsu, H. Pathophysiologic role of histamine: evidence clarified by histidine decarboxylase gene knockout mice. *Int. Arch. Allergy Immunol.,* **2012**, *158*(1) Suppl. 1, 2-6.
[http://dx.doi.org/10.1159/000337735] [PMID: 22627359]

[101] Takei, S.; Shimago, A.; Iwashita, M.; Kumamoto, T.; Kamuro, K.; Miyata, K. Urinary N-methylhistamine in asthmatic children receiving azelastine hydrochloride. *Ann. Allergy Asthma Immunol.,* **1997**, *78*(5), 492-496.
[http://dx.doi.org/10.1016/S1081-1206(10)63237-1] [PMID: 9164363]

[102] Hasegawa, S.; Ichiyama, T.; Sonaka, I.; Ohsaki, A.; Okada, S.; Wakiguchi, H.; Kudo, K.; Kittaka, S.; Hara, M.; Furukawa, S. Cysteine, histidine and glycine exhibit anti-inflammatory effects in human coronary arterial endothelial cells. *Clin. Exp. Immunol.,* **2012**, *167*(2), 269-274.

[http://dx.doi.org/10.1111/j.1365-2249.2011.04519.x] [PMID: 22236003]

[103] Alderton, W.K.; Cooper, C.E.; Knowles, R.G. Nitric oxide synthases: structure, function and inhibition. *Biochem. J.,* **2001**, *357*(Pt 3), 593-615.
[http://dx.doi.org/10.1042/bj3570593] [PMID: 11463332]

[104] Morris, C.R.; Poljakovic, M.; Lavrisha, L.; Machado, L.; Kuypers, F.A.; Morris, S.M., Jr Decreased arginine bioavailability and increased serum arginase activity in asthma. *Am. J. Respir. Crit. Care Med.,* **2004**, *170*(2), 148-153.
[http://dx.doi.org/10.1164/rccm.200309-1304OC] [PMID: 15070820]

[105] Böger, R.H.; Vallance, P.; Cooke, J.P. Asymmetric dimethylarginine (ADMA): a key regulator of nitric oxide synthase. *Atheroscler. Suppl.,* **2003**, *4*(4), 1-3.
[http://dx.doi.org/10.1016/S1567-5688(03)00027-8] [PMID: 14664896]

[106] Tapiero, H.; Mathé, G.; Couvreur, P.; Tew, K.D., II II. Glutamine and glutamate. *Biomed. Pharmacother.,* **2002**, *56*(9), 446-457.
[http://dx.doi.org/10.1016/S0753-3322(02)00285-8] [PMID: 12481981]

[107] Luo, Y.; Zhu, J.; Gao, Y. Metabolomic analysis of the plasma of patients with high-altitude pulmonary edema (HAPE) using 1H NMR. *Mol. Biosyst.,* **2012**, *8*(6), 1783-1788.
[http://dx.doi.org/10.1039/c2mb25044f] [PMID: 22498880]

[108] Pouw, E.M.; Schols, A.M.; Deutz, N.E.; Wouters, E.F. Plasma and muscle amino acid levels in relation to resting energy expenditure and inflammation in stable chronic obstructive pulmonary disease. *Am. J. Respir. Crit. Care Med.,* **1998**, *158*(3), 797-801.
[http://dx.doi.org/10.1164/ajrccm.158.3.9708097] [PMID: 9731007]

[109] Morrison, W.L.; Gibson, J.N.A.; Scrimgeour, C.; Rennie, M.J. Muscle wasting in emphysema. *Clin. Sci. (Lond.),* **1988**, *75*(4), 415-420.
[http://dx.doi.org/10.1042/cs0750415] [PMID: 3197374]

[110] Meléndez-Hevia, E.; De Paz-Lugo, P.; Cornish-Bowden, A.; Cárdenas, M.L. A weak link in metabolism: the metabolic capacity for glycine biosynthesis does not satisfy the need for collagen synthesis. *J. Biosci.,* **2009**, *34*(6), 853-872.
[http://dx.doi.org/10.1007/s12038-009-0100-9] [PMID: 20093739]

[111] Gebicki, J.; Sysa-Jedrzejowska, A.; Adamus, J.; Woźniacka, A.; Rybak, M.; Zielonka, J. 1-Methylnicotinamide: a potent anti-inflammatory agent of vitamin origin. *Pol. J. Pharmacol.,* **2003**, *55*(1), 109-112.
[PMID: 12856834]

[112] Wittmann, C.; Krömer, J.O.; Kiefer, P.; Binz, T.; Heinzle, E. Impact of the cold shock phenomenon on quantification of intracellular metabolites in bacteria. *Anal. Biochem.,* **2004**, *327*(1), 135-139.
[http://dx.doi.org/10.1016/j.ab.2004.01.002] [PMID: 15033521]

[113] O'Hagan, S.; Dunn, W.B.; Knowles, J.D.; Broadhurst, D.; Williams, R.; Ashworth, J.J.; Cameron, M.; Kell, D.B. Closed-loop, multiobjective optimization of two-dimensional gas chromatography/mass spectrometry for serum metabolomics. *Anal. Chem.,* **2007**, *79*(2), 464-476.
[http://dx.doi.org/10.1021/ac061443+] [PMID: 17222009]

[114] Rasmussen, L.G.; Savorani, F.; Larsen, T.M.; Dragsted, L.O.; Astrup, A.; Engelsen, S-B. Standardization of factors that influence human urine metabolomics. *Metabolomics,* **2011**, *7*, 71-83.
[http://dx.doi.org/10.1007/s11306-010-0234-7]

[115] Motta, A.; Paris, D.; Melck, D.; de Laurentiis, G.; Maniscalco, M.; Sofia, M.; Montuschi, P. Nuclear magnetic resonance-based metabolomics of exhaled breath condensate: methodological aspects. *Eur. Respir. J.,* **2012**, *39*(2), 498-500.
[http://dx.doi.org/10.1183/09031936.00036411] [PMID: 22298616]

[116] Ogata, H.; Goto, S.; Sato, K.; Fujibuchi, W.; Bono, H.; Kanehisa, M. KEGG: Kyoto Encyclopedia of Genes and Genomes. *Nucleic Acids Res.,* **1999**, *27*(1), 29-34.

[117] *Kyoto Encyclopedia of Genes and Genomes.*, www.genome.jp/kegg

[118] Wishart, D.S.; Tzur, D.; Knox, C.; Eisner, R.; Guo, A.C.; Young, N.; Cheng, D. Jewell.; K, Arndt, D.; Sawhney, S.; Fung, C.; Nikolai, L.; Lewis, M.; Coutouly, M.A.; Forsythe, I.; Tang, P.; Shrivastava, S.; Jeroncic, K.; Stothard, P.; Amegbey, G.; Block, D.; Hau, D.D.; Wagner, J.; Miniaci, J.; Clements, M.; Gebremedhin, M.; Guo, N.; Zhang, Y.; Duggan, G.E.; Macinnis, G.D.; Weljie, A.M.; Dowlatabadi, R.; Bamforth, F.; Clive, D.; Greiner, R.; Li, L.; Marrie, T.; Sykes, B.D.; Vogel, H.J.; Querengesser, L. HMDB: the Human Metabolome Database. *Nucleic Acids Res.,* **2007**, *35*(1), 521-526. [http://dx.doi.org/10.1093/nar/gkl923]

[119] *Human Metabolome Database.*, www.hmdb.ca

[120] *Kyoto Encyclopedia of Genes and Genomes. Diseases,*   www. kegg. jp/ dbget - bin/www_bget?ds:H00079

CHAPTER 5

Decoding DNA Structure using NMR Spectroscopy

Mahima Kaushik[1,2,*], Swati Chaudhary[2], Sonia Khurana[2], Komal Mehra[2] and Shrikant Kukreti[2]

[1] *Cluster Innovation Centre, University of Delhi, Delhi, India*

[2] *Nucleic Acids Research Laboratory, Department of Chemistry, University of Delhi, Delhi, India*

Abstract: During the past few decades, Nuclear Magnetic Resonance (NMR) spectroscopy has been extensively used for decoding the nucleic acid structure. The presence of nuclei possessing net spin (^{1}H, ^{13}C, ^{15}N and ^{31}P) in the basic unit of DNA makes it a befitting subject to be studied, utilizing the tenets of NMR spectroscopy. Apart from elucidating the structure, magnetic resonance spectroscopy also uncovers the strand multiplicity and thereby successfully differentiates among various secondary structures of DNA, for instance, duplex, hairpin, triplex, i-motif and quadruplex.

NMR spectroscopy also unravels interactions of nucleic acids with ligands like drugs, mutagens, and proteins. The highlighting feature in elucidating the structure and dynamics of DNA interaction with ligands is that these studies can be conducted in their natural solution environment. The interpretation of structural and chemical basis of ligands is very crucial for the development of new therapeutic agents. NMR parameters like coupling constant and peak integration successfully shed light on integral features of DNA structure such as glycosidic bond angles, sugar pucker conformations and dihedral angles. Other geometrical properties including bent helices, coaxial stacking and non-Watson-Crick base pairing can also be explored using NMR spectroscopy.

This chapter aims to provide a paradigm to understand the features of ^{1}H, ^{31}P, ^{13}C NMR spectroscopy involved in the determination of nucleic acid structure. It also outlines the characteristic features of NMR spectra, which are associated with various DNA topologies.

Keywords: Applications of NMR, NMR of DNA, NMR of DNA triplex, NMR of Duplex, NMR of Quadruplex, NMR Spectroscopy.

INTRODUCTION

Deoxyribonucleic acid (DNA) is an extremely important biomolecule with a very complex structure. It contains important genetic information for various life

---

[*] **Corresponding author Mahima Kaushik:** Cluster Innovation Centre, University of Delhi, Delhi, India; Tel: 91-11-27666702; Emails: mkaushik@cic.du.ac.in; kaushikmahima@yahoo.com

Atta-ur-Rahman & M. Iqbal Choudhary (Eds.)

processes. It is essential to decode the structural features of DNA, as it transcribes to form RNA, which ultimately translates into proteins. Apart from the classical A-, B-, Z- form of DNA, it can also adopt various polymorphic structures like hairpin, triplex, i-motif [1] and G-quadruplex [2] depending upon various conditions like sequence of oligonucleotide, pH and salt [3]. These polymorphic structures can be characterized by a number of spectroscopic techniques including NMR, Circular Dichroism (CD) spectroscopy, and UV-thermal melting. CD spectroscopy is a highly sensitive technique to explore the secondary structure and global conformation of DNA. CD spectroscopy works on the principle of absorption of plane polarized light by a chiral molecule and the difference in absorption of right and left circularly polarized light components is termed as ellipticity. The resultant CD curve of ellipticity as a function of wavelength is characteristic for a particular molecule. The CD curve for B-form of DNA gives a positive peak in the range of 260-280 nm and a negative peak at 245 nm, which correspond to base stacking and helicity, respectively. For triplex-DNA, positive band appears at 280 nm and negative bands appear at 212 nm and 245 nm. For parallel G-quadruplex, positive bands appear at two different wavelengths *i.e.* 210 and 260 nm, whereas negative band appears at 240 nm. On the other hand, anti-parallel G-quadruplex exhibits positive bands at 210 and 290 nm along with a negative band at 260 nm [4, 5]

UV-Thermal melting studies can also be exploited to investigate the thermal stability of DNA in specific solution conditions. Different forms of DNA exhibit characteristic thermal melting profiles depending on the bonding and interactions present in DNA structure. Thermal melting of DNA structures is usually recorded at 260 nm. G-quadruplex gives characteristic inverted thermal melting profile at 295 nm [6]. Intramolecular DNA triple helix can also be characterized using T_M analysis. The parallel triplex is less stable than the anti-parallel triplex and it melts in a biphasic manner. The structure of parallel triplex varies by changing the salt (KCl) concentration, resulting in an alteration in thermal melting temperature but no change is observed in case of antiparallel triplex [7].

NMR is one of the most important tools to understand the structural aspects and dynamics of both proteins and nucleic acids [8]. NMR analysis has become quite easy with the recent advancements in computer technology and spectrometer instrumentation. NMR is found to be more advantageous over X-Ray crystallography because in later technique, molecules under investigation should be present in the crystal lattice state, whereas in NMR, one can work with molecules at natural conditions, without affecting their structural properties.

NMR spectroscopy utilizes radio frequency waves and can provide the detailed information about the magnetic nuclei. NMR Spectroscopy is based on the principle of nuclear magnetic resonance *i.e.* any atomic nuclei, which would be present in the magnetic fields will absorb electromagnetic radiation and then re-emit the same. Studies of nucleic acids using NMR methods have been expanded greatly in the last few years [9]. Initially, Kearns *et al.* in 1971, studied the NMR of transfer RNA (t-RNA), in order to examine the H-bonded proton resonance [10]. It was reported that the protons present on the nitrogen of the ring in uracil and guanine give the signals more downfield (~14 ppm), when they are involved in H-bonding using 2, 2-dimethyl-4-silapentane-1-sulphonate [11]. Later in 1982, first NMR spectra of DNA was published [12]. These results had opened up new vistas for the experimentalists through which these biological molecules like DNA, RNA and proteins could be observed. Further, a large number of experiments were performed for studying the stability and structure of nucleic acids. Earlier, NMR studies were confined to the characterization of only short DNA/ RNA oligonucleotides but with the technological improvements and supply of large amounts of long oligonucleotides, NMR has extended their limit to determine the 3-D structure of nucleic acids [13].

Various Types of NMR Utilized for the Elucidation of DNA Structure

Due to recent developments, nucleic acid structure can be decoded easily using different types of NMR like ^{1}H, ^{13}C, ^{15}N, and ^{31}P. A range of various types of NMR techniques have been summarized in Fig. (**1**). NMR of heteronuclear elements provide the sequence specific resonance assignment, spectral separation of proton resonance and facilitates the measurement of coupling constant. ^{31}P NMR is emanating as a powerful research tool to determine the structure and dynamics of a DNA molecule. Initially, ^{31}P NMR was restricted only for oligonucleotide sequences having length less than six base pairs but now it is applicable to even large DNA oligonucleotide stretches, as synthesis of large DNA sequence in high concentrations can be achieved with the help of new technologies. The root cause for assigning ^{31}P resonance of oligonucleotide sequence is to get the knowledge about the phosphodiester backbone conformation [14]. The linkage between nucleotides can be elucidated by the six torsional angles from one phosphate group to the succeeding one throughout the length of DNA backbone. ^{13}C NMR is an efficient, straightforward and convenient method of studying DNA deoxyribose ring pucker both in solid and solution state. Before 1989, ring conformation of deoxyribose was confirmed by the proton NMR spectra only, but now ^{13}C NMR is preferred over proton NMR, as even the small changes in the populations of ring conformers are expected to be easily detected by this technique [15]. Till now, very few ^{15}N NMR resonance spectra have been remarked [16], possibly due to low sensitivity of natural abundance of ^{15}N NMR spectroscopy, which leads to a remarkably broad resonance.

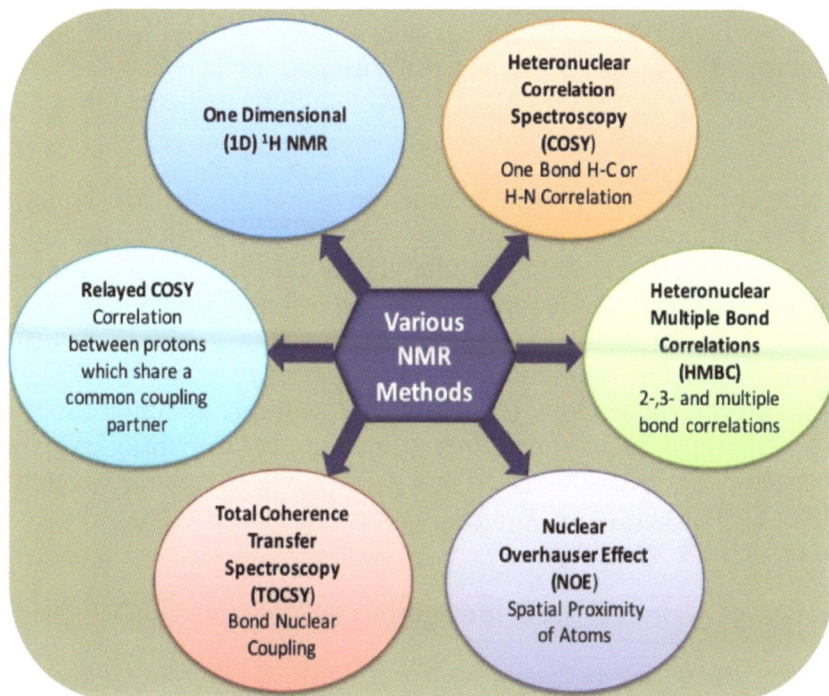

Fig. (1). Different types of NMR techniques.

NMR studies of DNA/ RNA are performed to investigate various factors like the distance between nuclei, coupling constant, torsion angles etc., which help in understanding the structural aspects of the molecule under consideration and its interaction with various ligands. For instance, the binding ability of Phenothiazines with different RNA elements was studied using NMR technique [17]. In order to examine DNA/ RNA by NMR spectroscopy, main focus is based on the proton and phosphorus nuclei, whereas more recent studies have devoted their efforts to carbon and nitrogen nuclei [4]. It would not be incorrect to say that the interaction and structural properties of biopolymers can be best studied using magnetic properties of nuclei having spin equals to half; such as ^{1}H, ^{13}C, ^{19}F and ^{31}P. Since natural nucleic acids do not contain any fluorine atom, so 2'-fluoro-2'-deoxyadenosine (a non-natural nucleotide) can be fused into the strands of nucleic acids, in order to use ^{19}F NMR technique [18].

Recent advancements in the NMR technique have extended the range of studies about various aspects of nucleic acids like their conformations and even dynamics [13]. NMR studies of the structure and dynamic aspects of nucleic acids are quite different from that of proteins. The NMR spectra of the DNA double helix become overcrowded due to stiffness and lesser number of hydrogen atoms and thereby it become much difficult to understand and interpret. Generally, two dimensional NMR (2D-NMR) is used in case of nucleic acids. NMR spectroscopy mainly includes the measurement of various parameters like the interproton distance by NOE (Nuclear Overhauser Effect), coupling constant (J) and chemical shifts.

In order to determine macromolecule structure by 2D-NMR, two steps should be followed. First is to assign the frequency label to every single proton, and the second is to estimate the strength of interaction between labeled protons. Spin-Spin coupling or J-coupling and dipole-dipole coupling are two main indirect and direct interactions, respectively, which are crucial in structure determination. Three main techniques namely Correlated Spectroscopy (COSY), Nuclear Overhauser Effect Spectroscopy (NOESY) and Multiple Quantum Spectroscopy (MQSY) are reported in literature to exploit the above mentioned parameters. From the spectra obtained, an analyst can gain knowledge about J-coupling from the multiplet fine structure peaks and interproton distance from the cross peak intensities in NOESY spectra [19].

NMR Studies of Various DNA Topologies

Nucleic acids have the potential to form various alternative structures by undergoing a phenomenon called as structural polymorphism. The variability of nucleic acid structures depends on several factors such as oligonucleotide sequence, it's concentration, pH, temperature and solvent conditions. Thus, NMR spectroscopists find the structural features of nucleic acids a fascinating subject to study. Characterization of DNA structure in solution state can be achieved by using NMR spectroscopy, which is very important as DNA performs its physiological function in the body fluids such as blood, saliva or stomach liquid [5]. A wide range of different types of DNA structures such as hairpin, duplex, triplex, i-motif and G-quadruplex have been very well studied and reviewed in a recent article from our own research group [3]. Various characteristics of all these structural polymorphs of DNA have been investigated using NMR spectroscopy. Protons involved in base pairing are considered as exchangeable protons and are characterized by NOESY technique, which establishes a correlation between neighboring bases. On the contrary, non-exchangeable protons or protons present in sugar moiety of DNA structure are identified using COSY and Total Correlated Spectroscopy (TOCSY) techniques. The correlation between a particular base to its neighboring base and the sugar to the base is done by NOESY experiments [20].

Relative coordinates of DNA duplex can be defined by NMR spectroscopy, due to which distinct H-bonding, ionic and Van der Waals interactions can be characterized. Different NMR techniques including 2D-NOESY, TOCSY and

DFQ-COSY (Double Quantum Filter-COSY) are used to deduce structural information consisting of torsional angles, dihedral angles and inter-proton distances [21]. Characteristic features of NMR spectra of various DNA topologies are discussed in the following sections.

DNA Duplex

Double helical structure is the first thing that comes to one's mind, while thinking about DNA. Double helical structures are usually acquired by the self-complementary DNA sequences, which are generally stabilized by the Watson-Crick base pairing and stacking interactions. Formation of double helical structure by self-complementary DNA oligonucleotides is favorable in two conditions; first when the negative charge of phosphate ions is neutralized by the counterions provided by salts and secondly, when DNA concentration is appropriately high enough so that every oligomer can find its complementary pair. Hence, it could be said that DNA concentration, salt concentration and temperature do have an extremely important role in the formation of DNA structures.

In NMR spectra, the regions 5.3-6.3 ppm and 7.1-8.4 ppm are considered as the fingerprint regions containing cross-peaks for H8/ H6 protons of nitrogenous bases and H1' protons of sugar residues. NMR chemical shift values of different nucleotide bases are given in Fig. (**2**).

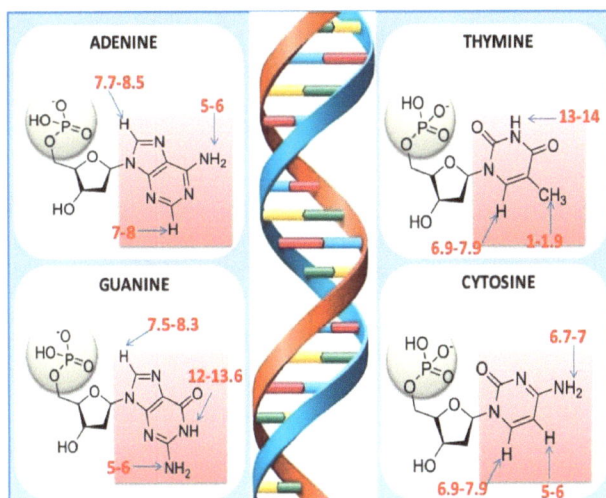

Fig. (2). Chemical shift values of different nucleotide bases present in DNA.

Different DNA conformations like B-form or A-form can also be distinguished by calculating volume of cross peaks. Sugar present in A-DNA is in the C3'-endo conformation, while sugar in B-DNA shows C2'-endo conformation. A-form of DNA is characterized by the larger cross peak volume, which indicates a smaller distance between the proton of nitrogenous base and sugar [22]. Two-types of peaks *i.e.*, intra-nucleotide as well as inter-nucleotide peaks are observed in this region. Z-DNA shows sensitive chemical shift values at 6.21 ppm for guanine H1' proton and 1.06 ppm for cytosine C5 CH3 proton [23]. Characteristic NMR shifts for B, A- as well as Z-DNA conformations are shown in Fig. (**3**). In 2002, Liu *et al.*, determined the structural features of the 17-mer DNA duplex by NMR spectroscopy and the samples were prepared in different ratios of H_2O: D_2O. NOESY spectrum at 10 $^{\circ}$C, which showed that fifteen out of seventeen imino-protons were interconnected to each other. Some of the imino-protons exhibit greater peak intensities than the others, whereas the peak of two imino-protons, which correspond to T-A base pairs disappears at 38 $^{\circ}$C. These NOE results further suggest that the conformation of the 17-mer duplex resembles B-form of DNA rather than A-DNA. Three base pairs at the end of the 17-mer duplex showed degenerate chemical shift values of the protons, which result from a symmetrical conformation, indicating a partially palindromic sequence [24].

Duplex DNA		
B-DNA	**A-DNA**	**Z-DNA**
Larger peak volume of base proton to H3' cross peak in comparison to H2'to base proton peak	Base proton to sugar H1' region (finger print region) 5.3-6.3 ppm and 7.1-8.4 ppm	Sensitive marker of Z-DNA- Guanine H1' proton at 6.21 ppm, Cytosine C-5 CH₃ group at 1.06 ppm, specific NOEs for Guanine H8 and H1' proton indicating *syn* glycosidic conformation

Fig. (3). Characteristic NMR peaks of B-, A- and Z-DNA.

Recently, structural features of the four sets of dodecamer base pairs were studied using NMR spectroscopy. 2D homonuclear NMR techniques such as COSY, NOESY and TOCSY were used to assign values to aromatic and ribose sugar protons of deoxy-oligonucleotide sequences. On the other hand, correlation between protons of aromatic and ribose regions was established by employing 2D ^{1}H-^{13}C HSQC and Transverse relaxation optimized spectroscopy (TROSY) experiments. With the help of different NMR techniques, huge data constituting 72 ^{31}P chemical shift values, 291 residual dipolar coupling (RDCs) values and 194 internucleotide distance (D_{inter}) values regarding the four DNA sequences under study was obtained. This huge NMR data was used to determine the relative distances and orientations between consecutive bases. It further established a relation between the backbone linkage conformations. The correlation between ^{31}P chemical shift values and the difference between two successive residual dipolar couplings (ΔRDCs) help in reflecting the population of two backbone conformation states in free B-DNA, which demonstrates the coordinated motions of phosphate groups and bases in free B-DNA in solution [25, 26].

DNA duplex generally adopts an antiparallel conformation but it also has an ability to adopt parallel conformation, which might play a crucial role in biological processes. Reverse Watson-Crick hydrogen bonding involved in A.T and G.C base pairs or Hoogsteen A.T or G.C$^+$ base pairs is known to stabilize the parallel duplex structures [27]. Formation of parallel duplexes take place *via* either homo-base pairing or hetero-base pairing [28]. Parallel duplexes containing homo-base pairs are very well characterized by different techniques, but structural characterization of parallel duplexes containing hetero-base pairs (mismatched or complementary base pairing) is still not done completely. In 2002, Parvathy *et al.*, used NMR spectroscopy to study the formation of parallel duplex containing complementary bases [29]. NOESY and E-COSY NMR techniques were used to study two dodecamer DNA oligonucleotides and the type of hydrogen bonding involved in the parallel duplex was also determined. NOE cross peaks for T(3NH)-A(H2) were observed, which suggested the formation of reverse Watson-Crick hydrogen bonding between A.T base pairs. NMR peaks of hydrogen-bonded imino-protons appear at a specified region, whereas non-hydrogen bonded imino protons appear upfield with broad peaks. NMR spectra showed that imino-protons of various purine and pyrimidine bases are involved in inter-base hydrogen bonding [29].

Hairpin DNA

In addition to the linear double helix formation, there is also a possibility of formation of 'hairpin' structures by DNA oligonucleotide sequences, which are usually formed by the self-complementary strands of DNA at a particular concentration of salt and oligomer. Hairpin structure holds two different parts- a double stranded long stem with Watson-crick base pairing and a single stranded loop, which might usually have 1 to 7 bases. DNA hairpin even with a single residue loop having quite strong stability had been reported from author's laboratory, which was shown to exhibit hairpin to duplex equilibrium [30].

Hairpin structures are present generally in naturally occurring RNA molecules like transfer-RNA, viroids, ribosomal-RNA and 5S RNA. The competition between hairpin structures and the Watson-Crick duplex makes hairpin less frequent in DNA. Under suitable conditions, the formation of hairpin structures plays an important biochemical role *in vitro* and *in vivo*. Studies involving the hybridization of DNA oligonucleotide sequence to its complementary sequence in molecular cloning can be affected by the formation of 'hairpin' structure by single stranded DNA, which contains complementary sequence within the same single strand. It has also been reported that d(CGCGTATACGCG) and d(CGCGAATTCGCG) self-complementary nucleotide sequences can even form a hairpin structure in solution [31]. For these many reasons, various studies involving the thermodynamics of hairpin formation, type of base pairing present, loop region structure and stability of hairpin with different loop sizes have continuously been conducted to investigate hairpin structures in synthetic DNA fragments.

Proton-NMR and other techniques were used by Satoshi Ikuta and her group members to explore the behavior of synthetic DNA oligonucleotide sequence d(CGCGCGTTTTCGCGCG) over a wide range of oligonucleotide concentration and temperature [31]. Various properties like thermal stability, loop structure and imino-proton exchange behavior of d(ATCCTATTTTATCC) and hexamer hairpin [d(CGCGCG)•d(CGCGCG)] had been compared.

High resolution NMR spectroscopy, commonly called as 2D-NMR has been evolved as a most promising and powerful tool for examining the structure of large biological molecules in solution state. NMR is more advantageous over other techniques to study the conformations that occur in solution, like hairpins. Various NMR techniques like COSY and NOESY enable the experimentalist to get the information about scalar couplings to get dihedral angle and distance for protons up to 4-5 Å apart respectively. From this information, one can deduce various biomolecular aspects like DNA conformation and secondary structure of proteins [32].

NMR spectroscopy is an extremely useful technique to elucidate the structure of biologically important materials like DNA, RNA and proteins. Chemical shift and spin-spin coupling constant values obtained from the NMR have quite close dependence on molecular geometry and hence can be used to identify structure, molecular conformation and chemical bonding analysis [33]. Information regarding the mutual interaction between nuclear spins of nuclei in pairs bonded by chemical bonds can be extracted by NMR spin-spin coupling constants, whereas the electronic environment in the locality of the nuclei under study can be monitored by NMR chemical shifts. Sychrovsky's group elucidated a DNA hairpin structure, using NMR [33]. They used the sequence d(G1C2G3A4A5G6C7) which forms DNA hairpin structure by utilizing the Watson-Crick base pairing between G1C7 and G6C2 and non-Watson-Crick base pairing between G3A5 along with an Adenine base A4 in loop.

Three decades ago, James R. Williamson and Steven G. Boxer did the comparative conformational studies of three related DNA hairpins using proton and phosphorous 2D-NMR technique [34]. The oligomer sequences studied were d(CGCGTTGTTCGCG) [hairpin 1], d(CGCGTTTGTCGCG) [hairpin 2], and d(CTGCTCTTGTTGAGCAG) [hairpin 3]. Each of these sequences formed hairpins with five bases in the loop. Hairpin 1 and 2 share the same stem region and these differ at the loop region, whereas the hairpin 2 and 3 have the same loop region but differ in stem sequence. From the experimental findings, it was concluded that the steric factor is superior in case of hairpin 1 and 2 and stem loop stacking prevails in case of hairpin 3. Obtained NOE patterns unveiled that the stacking arrangement in loop area relies on the base pairs that close the stem region.

Williamson and Boxer have also reported the hairpin conformation acquired by a DNA oligomeric sequence d(CGCGTTGTTCGCG) [35]. It contains five bases in the loop region and four C-G base pairs in the B-DNA like stem. Loop region of the hairpin structure also plays a very important role, as it resists a complete turn to permit base pairing of a single strand with itself. The oligomeric sequence was studied specifically by proton-NMR to get the knowledge about interproton distance and torsional angle.

NMR spectroscopy being a useful and informative technique contains a large amount of structural information in the form of interproton distance, which are not easily interpreted into 3-dimensional structure. The distance geometry mathematics was developed in 1930 and Crippen and Havel took the lead of its application to biological molecules [32]. J. J. Blommers and his group employed the two-dimensional spectroscopy to analyze the hairpin structure formed by the oligomeric sequence d(ATCCTATTTATAGGAT) [36]. The sequence was selected by the group to study, as it showed well-resolved proton NMR spectrum. Proton NMR and ^{31}P NMR peaks were assigned to the stem and loop region. The loop structure was derived in detail from NMR data, by utilizing some new and improved methods. Glycosidic torsion angle and pseudo rotational parameters were varied in order to gain the knowledge about the conformational space available to the selected oligonucleotide sequence.

DNA Triplex

In late 1950s and early 1960s, researchers had proposed that DNA can adopt a variety of unusual structures in addition to B-DNA *via* alternative hydrogen bonding arrangements under appropriate conditions. Certain molecular, biological and biochemical evidences demonstrated the formation of DNA triplexes and quadruplexes by the purine-rich sequences present in biological control regions of the genome [37]. Triple stranded DNA or DNA triplex is a DNA structure which is formed, when a third oligonucleotide strand binds to the major groove of the DNA double helix. The triplex forming oligonucleotide (TFO) binds to purine-rich strand of the DNA duplex *via* Hoogsteen hydrogen bonds, which provide the specificity and stability to the triplex structure. DNA triplexes are classified as intermolecular and intramolecular triplexes on the basis of the origin of the third strand *i.e.* whether third strand is provided by one of the strands of the same duplex DNA or by the different DNA molecule [38]. The pyrimidine•purine•pyrimidine triplex is well characterized, where pyrimidine motif binds to the purine strand of the duplex in parallel orientation *via* Hoogsteen hydrogen bonds. The Hoogsteen base pairing requires protonation of Cytosine at N3 and the resulting triplex comprised T•A•T and C+•G•C triplets. The purine•purine•pyrimidine triplex is unusually stable triplex, in which purine motif binds antiparallel to polypurine strand of Duplex DNA *via* reverse Hoogsteen hydrogen bonds. Multidimensional NMR spectra can form a suitable method for unravelling of such DNA triplexes at high resolution [37].

The presence of cations and pH play an important role in both, formation and the stability of DNA triplexes. Different spectra for DNA triplex are observed in the presence of different cations. Fig. (**4**) shows the formation of an intramolecular triplex and certain NMR signatures of base triplets. Generally, the optimal pH for the observation of imino-cross peaks in case of DNA triplexes is found to be 6 or lower. Since protonation of cytosine is required in case of pyrimidine motif triplexes, therefore, the optimal pH for observation of imino-cross peaks is found to be 5.2, so that equilibrium can be derived towards the triplex conformation [37].

One dimensional ^{1}H NMR spectra of a pyrimidine•purine•pyrimidine intramolecular triplex in water reveals that imino-protons of both Watson Crick and Hoogsteen base pairs resonate in the range of 12-16 ppm. The appearance of resonance peak of these additional Hoogsteen base pair imino-protons strongly recommends the triplex formation. In pyrimidine motif triplexes, the protonated carbon imino-protons resonate at lower field and peak appears between 14 to 16 ppm. Comparatively, low field resonance peak of protonated cytosine amino group appears in the range of 9 to 10.5 ppm than unprotonated cytosine amino proton, which also signifies the formation of DNA triplex [37].

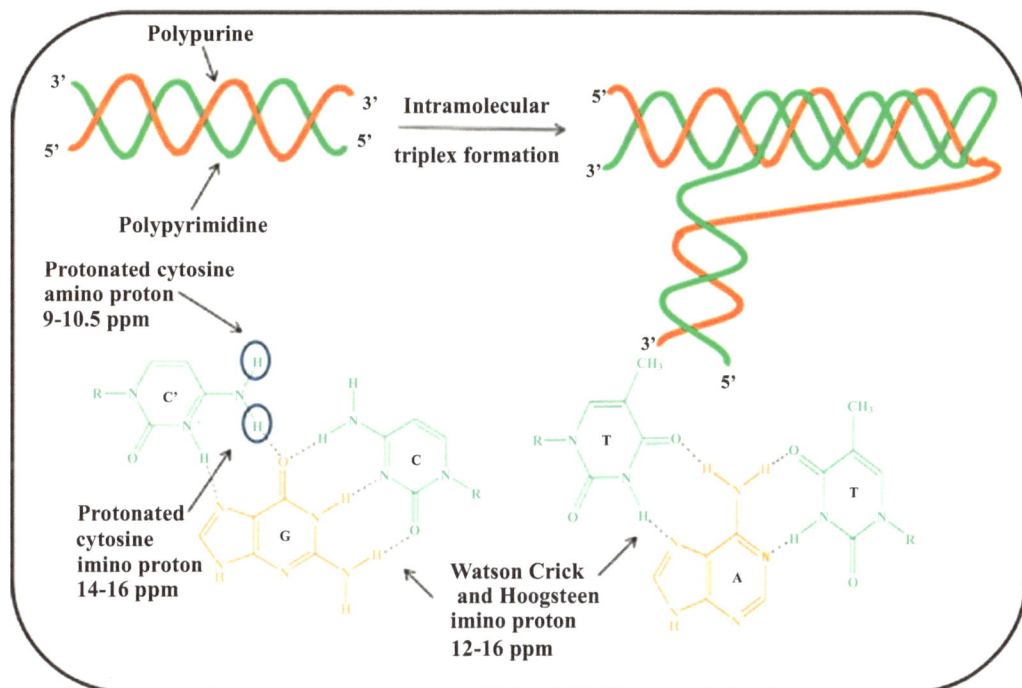

Fig. (4). NMR signatures of the base triplets of DNA triplex.

Shafer's group had investigated the structure of intermolecular purine•purine•pyrimidine DNA triplex with the help of various techniques including ^{1}H NMR spectroscopy [39]. One d(C$_3$T$_4$C$_3$) and two d(G$_3$A$_4$G$_3$) deoxyoligonucleotide sequences were associated to form triplex with Mg^{2+} counter ion at neutral pH. The triplex thus formed is stabilized by A•A•T and G•G•C base triplet and the formation of these new hydrogen bonds between purine bases in addition to classic Watson crick base pairs were characterized by recording proton NMR spectra of imino-protons. NMR spectrum of solution of oligonucleotides both in the presence as well as absence of counter ion Mg^{2+} was recorded. In the absence of counter ion, the ^{1}H NMR spectra of imino-protons is found to be identical to that of the DNA duplex, which indicates that the second purine strand was not forming triplex in these conditions and remains in unassociated single strand form. In the presence of Mg^{2+} counter ions, appearance of new imino-proton resonance occurs, postulating the formation of the G-G hydrogen bonds to form DNA triplex. These additional hydrogen bonds are between N7 atoms of guanine residues on the purine strand of the primary duplex and NH1 imino-proton of third strand guanine residue. The resonance line of these imino-protons in triplex appears because of their involvement in hydrogen bonding, due to which their exchange with the solvent becomes difficult. The hydrogen bonds between two adenine bases cannot be characterized by recording imino-proton resonances, due to the absence of imino-protons on adenine bases at neutral pH. But in the presence of Mg^{2+} ions, alterations in resonance peaks of imino-proton regions (6-10 ppm) occur, which signify the formation of A-A hydrogen bonds. The results from polyacrylamide gel electrophoresis (PAGE) showed evidence about the antiparallel orientation of the third strand with respect to the purine strand of the duplex [39].

Patel's group had studied the NMR spectra of intramolecular purine•purine•pyrimidine DNA triplex in aqueous solution [40]. A 31-mer deoxyoligonucleotide sequence was shown to fold intramolecularly, forming a 7-mer triplex stabilized by A•A•T and G•G•C base triplet. The third strand of DNA triplex was linked to the Watson-Crick strand in antiparallel fashion and the helical segments in triplex were connected by a loop size of five thymidine residues. The triplex was highly favored in the presence of Li^+ counter ions in aqueous solution compared to any other ion, as it facilitates its formation at the expense of G-quadruplex formation. The exchangeable and non-exchangeable proton spectra had been recorded in order to characterize the DNA triplex formed. The NOE parameters elucidate the structure of triplex such that segments (TCCCTCC) and (GGAGGGA) are aligned antiparallel through Watson-Crick base pairing, whereas (TGGGTGG) is also aligned antiparallel to purine strand of the duplex through Hoogsteen base pairing. Despite of low temperature, the exchangeable protons resonate in the range of 8.4-14.5 ppm, whereas non-exchangeable protons resonate between 5.5-8.0 ppm. Among exchangeable protons, hydrogen-bonded imino-protons resonate between 12.3-14.5 ppm, whereas imino-protons in loop segments resonate between 10-11.2 ppm and hydrogen-bonded cytidine amino protons resonate between 8.4 and 9.2 ppm. Like duplex DNA, the imino-protons of thymidine residues in triplex resonate at relatively downfield *i.e.* 14.2 to 14.5 ppm than imino-protons of guanosine residues, which resonate in the range of 12.8 to 13 ppm. The hydrogen bonded amino protons of cytidine residues of the triplex resonate in the range of 8.4 to 9.2 ppm, which is relatively downfield than their counterparts involved in G•C base-pairing in DNA duplex resonating between 8.0 to 8.5 ppm. The in-plane ring current contributions are from the guanine residues of the third strand of DNA triplex, while the formation of G•GC triplets is assumed to be responsible for the downfield shift. Unlike imino-protons of cytidine residues, the imino-protons of thymidine residues engaged in Hoogsteen base pairing resonate between 12.3 to 13.2 ppm, which is up field relative to the thymidine residues forming Watson-Crick base pairing in duplex DNA. The imino-protons of guanosine residues involved in Watson-Crick or Hoogsteen base pairing resonate in the similar range *i.e.* 12.5 to 13.8 ppm. The slight downfield chemical shift of third strand guanosine imino-protons suggest that the nitrogen atom is involved in hydrogen bonding rather than oxygen atom. The hydrogen bonding in G•GC triplets take place between the N-7 ring atom of guanosine on the purine strand of the underlying duplex and the imino-proton of guanosine on the third strand. The base triplet pairing alignments and strand directions in DNA triplex can be better understood by observing NOE cross peaks of imino-protons [40].

Guanine-Quadruplex

Several types of polymorphic guanine-rich DNA structures having different glycosidic conformations, strand orientation, groove width, and loop arrangements have been well documented, commonly known as G-quadruplex structures. Sequence of the G-rich oligonucleotide, nature of counter ions, strand concentrations, and solution conditions are the factors on which the structural polymorphism of quadruplex depends. Various techniques such as gel electrophoresis, circular dichroism (CD), mass spectrometry, UV spectroscopy and chromatography have been employed to determine the structural features of G-quadruplexes [5]. High resolution of G-quadruplex structures in solution can be further determined by NMR spectroscopy, which also simultaneously investigates the molecular interaction as well as the dynamic and kinetic studies [41]. The guanine imino-protons (H1) present in G-quartet show their chemical shift peaks in the range of 10-12 ppm, whereas imino-protons involved in Watson-Crick base pairing exhibit their NMR peaks within the range of 13-14 ppm. In G-quadruplex structures, the rate of exchange of guanine imino-protons with solvent is very slow as compared to exchangeable protons in Watson-Crick duplex [42]. NMR spectroscopy helps in diagnosing the different G-quadruplex conformations formed by a single G-rich sequence by counting the exceeding NMR peaks of guanine imino-protons. The stoichiometry of G-quadruplex structures can be indirectly deduced from the NOE restraints and spectral line-width data. In order to determine the stoichiometry directly from NMR techniques, NMR spectra of G-quadruplex was monitored at different strand concentrations [43].

In order to examine NMR spectra, assigning resonance peaks is the most crucial step, which is conventionally done by using NOE and through bond correlations. A.T. Phan in 2002 showed the site specific, N^{15}, C^{13} isotope labeling method to assign NMR resonance peaks on the dimeric G-quadruplex of DNA sequence d(GGGTTCAGG) [44]. The G-rich oligonucleotide sequences (GGGT)$_4$ and

(GGGC)$_4$ named *T30695* and *T40214* respectively, are known to show anticancer as well as anti-HIV activity [45]. Their structural features were characterized by NMR spectroscopy techniques, including 1D NMR and 2D NOESY. NMR spectra indicated that both the sequences form similar kind of G-quadruplex structures. Strong intra-residue H8-H1' NOE cross-peaks for guanines were not observed in the case of (GGGT)$_4$ sequence, which indicates that syn glycosidic conformation is not involved in the G-quadruplex structure. The NMR peaks of *T30695* are found to be highly overlapped, due to which the structural analysis becomes difficult. In order to improve the NMR spectra, the guanine residue at position 2 of *T30695* was replaced by inosine. As a result of which, sharp NMR peaks at 10.8-11.6 ppm were observed for eleven guanine imino-protons and one NMR peak at 13.8 ppm was also observed for inosine imino-proton, which corresponds to three G-tetrads [45]. The structure of G-quadruplex containing 15 continuous guanine residues was very well characterized by NMR spectrum. The NMR spectra displayed twelve sharp peaks and indicate the formation of propeller-type parallel G-quadruplex structure, containing three layers of G-quartet.

The 2D NOESY spectra of G15 revealed that all the guanines adopt *anti*-glycosidic conformation [46]. Recently, important features of the G-quadruplex structure formed by KRAS promoter region are revealed by high-resolution 3D NMR spectroscopy. Specific intra-quartet NOE correlations show that three G-tetrads are involved in the folding pattern to give monomeric G-quadruplex structure. The imino-protons of the guanines forming central tetrad are found to be protected from water/ deuterium exchange as compared to other guanine residues. The NOE inter-residue connectivity data revealed that the protons of guanines of central tetrad are protected by the single nucleotide chain reversal loop, which completely blocks the groove area of this central tetrad. All the glycosidic torsion angles have adopted *anti*-conformation as suggested by intra-guanine NOE cross-peaks. NMR data in combination of other spectroscopic techniques indicate the formation of parallel G-quadruplex that plays an important role in transcriptional regulation [46]. Fig. (5). depicts the NMR signatures of different topologies of G-quadruplex structures [47 - 49].

The formation of more than one G-quadruplex structure in solution can be determined by diffusion ordered NMR spectroscopy (DOSY). This NMR technique is used to identify different macromolecules according to their molecular weight present in solution. DOSY-NMR depends on the shape, solution conditions as well as relaxation characteristics of the molecule. The stoichiometry of multistranded DNA conformations can be deduced by DOSY-NMR technique [50].

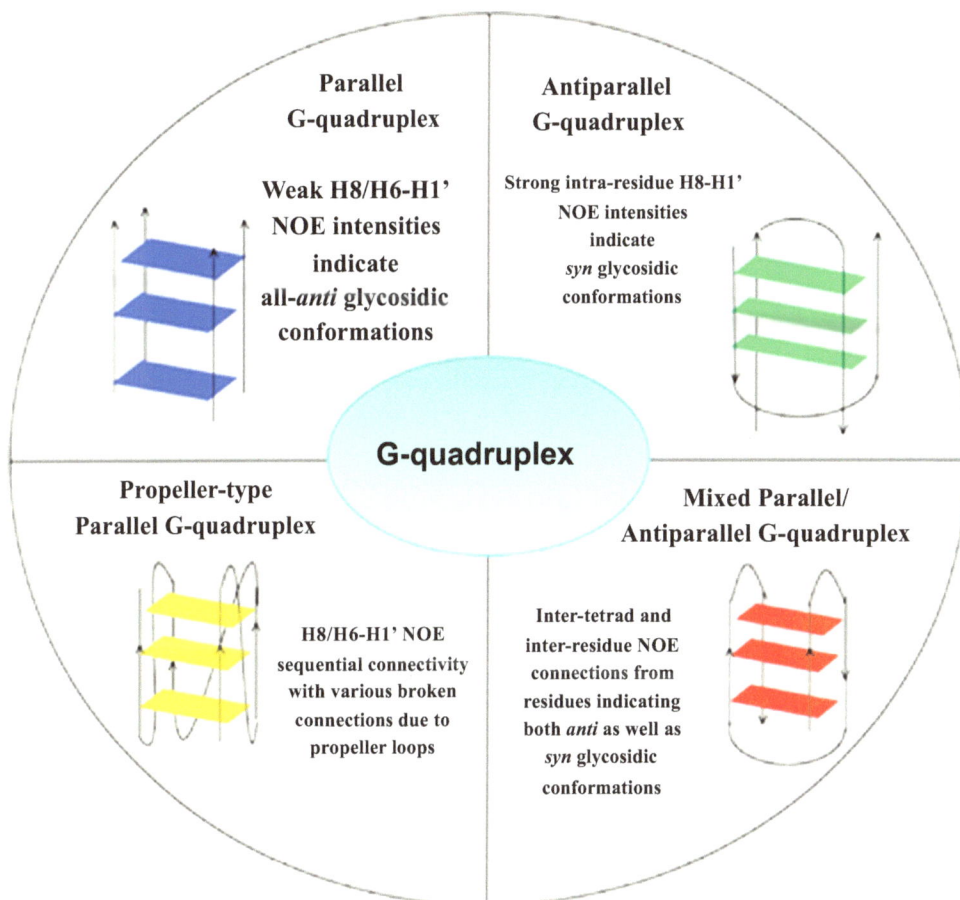

Fig. (5). NMR signatures of different topologies of G-quadruplex structure.

OUTLOOK AND FUTURE ASPECTS

NMR spectroscopy has been a fascinating tool for the investigation of complex biological molecules, for quite some time now. In addition, NMR also helps in structure elucidation and identification of chemical compounds, their quantitative analysis, quality assurance and investigating reaction mechanisms. Likewise, NMR spectroscopy can be utilized to obtain information related to structure and dynamics of nucleic acids such as DNA or RNA. It is proved useful for nucleic acids containing up to even 100 nucleotides. To improve the resolution of highly complex NMR spectra, two-dimensional NMR has been introduced. It has gained popularity over the X-ray crystallographic technique by determining the dynamics of DNA in crystal lattice instead of the solution state. Various alternative DNA structures such as duplex, hairpin, triplex, i-motif and quadruplex structures can also be well studied using this technique. The stability as well as intermolecular interactions of these DNA structures and DNA-ligand interactions can also be investigated by NMR. The transverse relaxation-optimized spectroscopy (TROSY) has been developed to attain cross-correlation between the chemical shift anisotropy and dipole-dipole interactions in order to enhance sensitivity and resolution. NMR spectroscopy along with X-ray crystallography has really been instrumental in increasing the knowledge of establishing a relationship between structure and function of biomolecules like nucleic acids, proteins and carbohydrates. Advancements in instrumentation and computational approaches are still needed for extending the repertoire of NMR studies of biomolecules.

CONSENT FOR PUBLICATION

Not applicable.

CONFLICT OF INTEREST

The editor declares no conflict of interest, financial or otherwise.

ACKNOWLEDGEMENTS

Mk would like to acknowledge the research and development grant (2015-16) of University of Delhi, Delhi, India.

REFERENCES

[1]　Ahmed, S.; Kaushik, M.; Chaudhary, S.; Kukreti, S. Structural polymorphism of a cytosine-rich DNA sequence forming i-motif structure: Exploring pH based biosensors. *Int. J. Biol. Macromol.,* **2018**, *111*, 455-461.
[http://dx.doi.org/10.1016/j.ijbiomac.2018.01.053] [PMID: 29329816]

[2]　Kaushik, M.; Kaushik, S.; Kukreti, S. Advancement in the structural polymorphism of G-quadruplexes. *International review of Biophysical Chemistry, 5,* **2014**.

[3]　Kaushik, M.; Kaushik, S.; Roy, K.; Singh, A.; Mahendru, S.; Kumar, M.; Chaudhary, S.; Ahmed, S.; Kukreti, S. A bouquet of DNA structures: Emerging diversity. *Biochem. Biophys. Rep.,* **2016**, *5*, 388-395.
[http://dx.doi.org/10.1016/j.bbrep.2016.01.013] [PMID: 28955846]

[4]　Kypr, J.; Kejnovská, I.; Renčiuk, D.; Vorlíčková, M. Circular dichroism and conformational polymorphism of DNA. *Nucleic Acids Res.,* **2009**, *37*(6), 1713-1725.
[http://dx.doi.org/10.1093/nar/gkp026] [PMID: 19190094]

[5]　Kaushik, M.; Kaushik, S.; Kukreti, S. Exploring the characterization tools of Guanine-Quadruplexes. *Front. Biosci.,* **2016**, *21*, 468-478.
[http://dx.doi.org/10.2741/4402] [PMID: 26709787]

[6]　Mergny, J.L.; Lacroix, L. UV Melting of G-Quadruplexes. *Curr. Protoc. Nucleic Acid Chem.,* **2009**, *Chapter 17*, 1.
[PMID: 19488970]

[7]　Gondeau, C.; Maurizot, J.C.; Durand, M. Circular dichroism and UV melting studies on formation of an intramolecular triplex containing parallel T*A:T and G*G:C triplets: netropsin complexation with

the triplex. *Nucleic Acids Res.,* **1998**, *26*(21), 4996-5003.
[http://dx.doi.org/10.1093/nar/26.21.4996] [PMID: 9776765]

[8]   Hare, D.R.; Wemmer, D.E.; Chou, S.H.; Drobny, G.; Reid, B.R. Assignment of the non-exchangeable proton resonances of d(C-G-C-G-A-A-T-T-C-G-C-G) using two-dimensional nuclear magnetic resonance methods. *J. Mol. Biol.,* **1983**, *171*(3), 319-336.
[http://dx.doi.org/10.1016/0022-2836(83)90096-7] [PMID: 6317867]

[9]   Zídek, L.; Štefl, R.; Sklenár, V. NMR methodology for the study of nucleic acids. *Curr. Opin. Struct. Biol.,* **2001**, *11*(3), 275-281.
[http://dx.doi.org/10.1016/S0959-440X(00)00218-9] [PMID: 11406374]

[10]   Kearns, D.R.; Patel, D.J.; Shulman, R.G. High resolution nuclear magnetic resonance studies of hydrogen bonded protons of tRNA in water. *Nature,* **1971**, *229*(5283), 338-339.
[http://dx.doi.org/10.1038/229338a0] [PMID: 4927207]

[11]   Shulman, R., Ed. *Biological applications of magnetic resonance*; Elsevier, **2012**.

[12]   Kenneth, J. "Introduction to 1H NMR Spectroscopy of DNA". In: *Bioorganic Chemistry: Nucleic Acids*; Oxford University Press: New York, **1996**.

[13]   Patel, D.J.; Shapiro, L.; Hare, D. DNA and RNA: NMR studies of conformations and dynamics in solution. *Q. Rev. Biophys.,* **1987**, *20*(1-2), 35-112.
[http://dx.doi.org/10.1017/S0033583500004224] [PMID: 2448843]

[14]   Gorenstein, D.G. ^{31}P NMR of DNA. *Methods Enzymol.,* **1992**, *211*, 254-286.
[http://dx.doi.org/10.1016/0076-6879(92)11016-C] [PMID: 1406310]

[15]   Braun, W.; Wagner, G.; Worgotter, E.; Vasak, M.; Kagi, J.; Clore, G. M. Scheek, R. M. Determination of the DNA sugar pucker using 13C NMR Spectroscopy. *Biol, 125*, 357-386.

[16]   James, T.L.; James, J.L.; Lapidot, A. Structural and dynamic information about double-stranded DNA from nitrogen-15 NMR spectroscopy. *J. Am. Chem. Soc.,* **1981**, *103*(22), 6748-6750.
[http://dx.doi.org/10.1021/ja00412a039]

[17]   Mayer, M.; James, T.L. NMR-based characterization of phenothiazines as a RNA binding scaffold. *J. Am. Chem. Soc.,* **2004**, *126*(13), 4453-4460.
[http://dx.doi.org/10.1021/ja0398870] [PMID: 15053636]

[18]   Kan, Lou-sing; Ts'o, R.V. "Nuclear Magnetic Resonance studies of Nucleic acids" In: *NMR in Biology and Medicine.* ; Raven Press: New York, **1986**. ISBN 0-88167-231-9

[19]   Majumdar, A.; Hosur, R.V. Simulation of 2D NMR spectra for determination of solution conformations of nucleic acids. *Prog. Nucl. Magn. Reson. Spectrosc.,* **1992**, *24*(2), 109-158.
[http://dx.doi.org/10.1016/0079-6565(92)80007-3]

[20]   Soliva, R.; Monaco, V.; Gómez-Pinto, I.; Meeuwenoord, N.J.; Marel, G.A.; Boom, J.H.; González, C.; Orozco, M. Solution structure of a DNA duplex with a chiral alkyl phosphonate moiety. *Nucleic Acids Res.,* **2001**, *29*(14), 2973-2985.
[http://dx.doi.org/10.1093/nar/29.14.2973] [PMID: 11452022]

[21]   Gmeiner, W.H. NMR spectroscopy as a tool to investigate the structural basis of anticancer drugs. *Curr. Med. Chem.,* **1998**, *5*(2), 115-135.
[PMID: 9481037]

[22]   Pandav, K.; Pandya, P.; Barthwal, R.; Kumar, S. , **2012**. Structure Determination of DNA Duplexes by NMR.
[http://dx.doi.org/10.1007/978-3-642-23394-4_33]

[23]   Patel, D.J.; Kozlowski, S.A.; Nordheim, A.; Rich, A. Right-handed and left-handed DNA: studies of B- and Z-DNA by using proton nuclear Overhauser effect and P NMR. *Proc. Natl. Acad. Sci. USA,* **1982**, *79*(5), 1413-1417.
[http://dx.doi.org/10.1073/pnas.79.5.1413] [PMID: 6951185]

[24]   Liu, W.; Vu, H.M.; Kearns, D.R. 1 H NMR studies of a 17-mer DNA duplex. *Biochimica et Biophysica Acta (BBA)-. Gene Structure and Expression,* **2002,** *1574*(1), 93-99.
[http://dx.doi.org/10.1016/S0167-4781(01)00350-5]

[25]   Imeddourene, A.B.; Xu, X.; Zargarian, L.; Oguey, C.; Foloppe, N.; Mauffret, O.; Hartmann, B. The intrinsic mechanics of B-DNA in solution characterized by NMR. *Nucleic Acids Res.,* **2016,** *44*(7), 3432-3447.
[http://dx.doi.org/10.1093/nar/gkw084] [PMID: 26883628]

[26]   Vermeulen, A.; Zhou, H.; Pardi, A. Determining DNA global structure and DNA bending by application of NMR residual dipolar couplings. *J. Am. Chem. Soc.,* **2000,** *122*(40), 9638-9647.
[http://dx.doi.org/10.1021/ja001919l]

[27]   Szabat, M.; Pedzinski, T.; Czapik, T.; Kierzek, E.; Kierzek, R. Structural Aspects of the Antiparallel and Parallel Duplexes Formed by DNA, 2'-O-Methyl RNA and RNA Oligonucleotides. *PLoS One,* **2015,** *10*(11), e0143354.
[http://dx.doi.org/10.1371/journal.pone.0143354] [PMID: 26579720]

[28]   Jain, A.K.; Bhattacharya, S. Groove binding ligands for the interaction with parallel-stranded ps-duplex DNA and triplex DNA. *Bioconjug. Chem.,* **2010,** *21*(8), 1389-1403.
[http://dx.doi.org/10.1021/bc900247s] [PMID: 20509695]

[29]   Parvathy, V.R.; Bhaumik, S.R.; Chary, K.V.; Govil, G.; Liu, K.; Howard, F.B.; Miles, H.T. NMR structure of a parallel-stranded DNA duplex at atomic resolution. *Nucleic Acids Res.,* **2002,** *30*(7), 1500-1511.
[http://dx.doi.org/10.1093/nar/30.7.1500] [PMID: 11917010]

[30]   Kaushik, M.; Kukreti, R.; Grover, D.; Brahmachari, S.K.; Kukreti, S. Hairpin-duplex equilibrium reflected in the A-->B transition in an undecamer quasi-palindrome present in the locus control region of the human β-globin gene cluster. *Nucleic Acids Res.,* **2003,** *31*(23), 6904-6915.
[http://dx.doi.org/10.1093/nar/gkg887] [PMID: 14627823]

[31]   Ikuta, S.; Chattopadhyaya, R.; Ito, H.; Dickerson, R.E.; Kearns, D.R. NMR study of a synthetic DNA hairpin. *Biochemistry,* **1986,** *25*(17), 4840-4849.
[http://dx.doi.org/10.1021/bi00365a018] [PMID: 3768317]

[32]   Hare, D.R.; Reid, B.R. Three-dimensional structure of a DNA hairpin in solution: two-dimensional NMR studies and distance geometry calculations on d(CGCGTTTTCGCG). *Biochemistry,* **1986,** *25*(18), 5341-5350.
[http://dx.doi.org/10.1021/bi00366a053] [PMID: 3768352]

[33]   Sychrovský, V.; Vacek, J.; Hobza, P.; Žídek, L.; Sklenář, V.; Cremer, D. Exploring the Structure of a DNA Hairpin with the Help of NMR Spin-Spin Coupling Constants: An Experimental and Quantum Chemical Investigation. *J. Phys. Chem. B,* **2002,** *106*(39), 10242-10250.
[http://dx.doi.org/10.1021/jp020673z]

[34]   Williamson, J.R.; Boxer, S.G. Multinuclear NMR studies of DNA hairpins. 2. Sequence-dependent structural variations. *Biochemistry,* **1989,** *28*(7), 2831-2836.
[http://dx.doi.org/10.1021/bi00433a013] [PMID: 2742815]

[35]   Williamson, J.R.; Boxer, S.G. Multinuclear NMR studies of DNA hairpins. 1. Structure and dynamics of d(CGCGTTGTTCGCG). *Biochemistry,* **1989,** *28*(7), 2819-2831.
[http://dx.doi.org/10.1021/bi00433a012] [PMID: 2742814]

[36]   Blommers, M.J.; van de Ven, F.J.; van der Marel, G.A.; van Boom, J.H.; Hilbers, C.W. The three-dimensional structure of a DNA hairpin in solution two-dimensional NMR studies and structural analysis of d(ATCCTATTTATAGGAT). *Eur. J. Biochem.,* **1991,** *201*(1), 33-51.
[http://dx.doi.org/10.1111/j.1432-1033.1991.tb16253.x] [PMID: 1915376]

[37]   Feigon, J.; Koshlap, K.M.; Smith, F.W. 1H NMR spectroscopy of DNA triplexes and quadruplexes. *Methods Enzymol.,* **1995,** *261*, 225-255.

[http://dx.doi.org/10.1016/S0076-6879(95)61012-X] [PMID: 8569497]

[38]   Jain, A.; Wang, G.; Vasquez, K.M. DNA triple helices: biological consequences and therapeutic potential. *Biochimie,* **2008**, *90*(8), 1117-1130.
       [http://dx.doi.org/10.1016/j.biochi.2008.02.011] [PMID: 18331847]

[39]   Pilch, D.S.; Levenson, C.; Shafer, R.H. Structure, stability, and thermodynamics of a short intermolecular purine-purine-pyrimidine triple helix. *Biochemistry,* **1991**, *30*(25), 6081-6088.
       [http://dx.doi.org/10.1021/bi00239a001] [PMID: 2059618]

[40]   Radhakrishnan, I.; de los Santos, C.; Patel, D.J. Nuclear magnetic resonance structural studies of intramolecular purine.purine.pyrimidine DNA triplexes in solution. Base triple pairing alignments and strand direction. *J. Mol. Biol.,* **1991**, *221*(4), 1403-1418.
       [PMID: 1942059]

[41]   Webba da Silva, M. NMR methods for studying quadruplex nucleic acids. *Methods,* **2007**, *43*(4), 264-277.
       [http://dx.doi.org/10.1016/j.ymeth.2007.05.007] [PMID: 17967697]

[42]   Chung, W.J.; Heddi, B.; Schmitt, E.; Lim, K.W.; Mechulam, Y.; Phan, A.T. Structure of a left-handed DNA G-quadruplex. *Proc. Natl. Acad. Sci. USA,* **2015**, *112*(9), 2729-2733.
       [http://dx.doi.org/10.1073/pnas.1418718112] [PMID: 25695967]

[43]   Adrian, M.; Heddi, B.; Phan, A.T. NMR spectroscopy of G-quadruplexes. *Methods,* **2012**, *57*(1), 11-24.
       [http://dx.doi.org/10.1016/j.ymeth.2012.05.003] [PMID: 22633887]

[44]   Phan, A.T.; Patel, D.J. A site-specific low-enrichment (15)N,(13)C isotope-labeling approach to unambiguous NMR spectral assignments in nucleic acids. *J. Am. Chem. Soc.,* **2002**, *124*(7), 1160-1161.
       [http://dx.doi.org/10.1021/ja011977m] [PMID: 11841271]

[45]   Do, N.Q.; Lim, K.W.; Teo, M.H.; Heddi, B.; Phan, A.T. Stacking of G-quadruplexes: NMR structure of a G-rich oligonucleotide with potential anti-HIV and anticancer activity. *Nucleic Acids Res.,* **2011**, *39*(21), 9448-9457.
       [http://dx.doi.org/10.1093/nar/gkr539] [PMID: 21840903]

[46]   Sengar, A.; Heddi, B.; Phan, A.T. Formation of G-quadruplexes in poly-G sequences: structure of a propeller-type parallel-stranded G-quadruplex formed by a G_{15} stretch. *Biochemistry,* **2014**, *53*(49), 7718-7723.
       [http://dx.doi.org/10.1021/bi500990v] [PMID: 25375976]

[47]   Kerkour, A.; Marquevielle, J.; Ivashchenko, S.; Yatsunyk, L.A.; Mergny, J.L.; Salgado, G.F. High-resolution 3D NMR structure of the KRAS proto-oncogene promoter reveals key features of a G-quadruplex involved in transcriptional regulation. *J. Biol. Chem.,* **2017**, *292*, 8082-8091.
       [http://dx.doi.org/10.1074/jbc.M117.781906] [PMID: 28330874]

[48]   Do, N.Q.; Chung, W.J.; Truong, T.H.A.; Heddi, B.; Phan, A.T. G-quadruplex structure of an anti-proliferative DNA sequence. *Nucleic Acids Res.,* **2017**, *45*(12), 7487-7493.
       [http://dx.doi.org/10.1093/nar/gkx274] [PMID: 28549181]

[49]   Ambrus, A.; Chen, D.; Dai, J.; Bialis, T.; Jones, R.A.; Yang, D. Human telomeric sequence forms a hybrid-type intramolecular G-quadruplex structure with mixed parallel/antiparallel strands in potassium solution. *Nucleic Acids Res.,* **2006**, *34*(9), 2723-2735.
       [http://dx.doi.org/10.1093/nar/gkl348] [PMID: 16714449]

[50]   Ambrus, A.; Yang, D. Diffusion-ordered nuclear magnetic resonance spectroscopy for analysis of DNA secondary structural elements. *Anal. Biochem.,* **2007**, *367*(1), 56-67.
       [http://dx.doi.org/10.1016/j.ab.2007.04.025] [PMID: 17570331]

CHAPTER 6

Early Diagnosis of Cancer using Nuclear Magnetic Resonance Spectroscopy: A Novel Diagnostic Approach

Neetu Talreja[1], Manjula Nair[2] and **Dinesh Kumar[3],***

[1] *Department of Bio-nanotechnology, Gachon University, Gyeonggi-do, South Korea*

[2] *HBMSU, Academic City, Dubai, UAE*

[3] *School of Chemical Sciences, Central University of Gujrat, Gandhinagar 382030, India*

Abstract: Cancer is an abnormal growth of cells in the body that spread through the organs and leads to health complications and sometimes, death. The most important challenge is to develop strategies for diagnosing and monitoring the risk of cancer, thereby effectively treating cancer patients.

Mutations in cancer genes and alterations in signals from cells might trigger a change in the metabolism. Metabolites represent the end products of complex metabolic pathways. The metabolome reflects changes by cancerous cells in cell cycle pathways, thereby providing a logical approach for cancer diagnosis. Specific changes in the metabolome are thought to reflect pathological states of patients. Based on the grade of degeneration, tumor cells show alterations in basic biochemical processes. The metabolic signature of malignancy and precursor cells in cancer metabolic reaction might provide an indication for the presence of cancer. In general, measuring the metabolites in cancer patients might be an effective strategy for early diagnosis of cancer.

Nuclear magnetic resonance (NMR) is a promising method for measuring concentrations of metabolites in complex samples with good reproducibility, high sensitivity, and simple sample processing. A positive aspect of NMR is that samples are not destroyed by the process; hence they can be analyzed in other ways too. In this chapter, we summarize the uses of NMR spectroscopy in early diagnosis of cancer diseases as well as future prospects of this technique.

Keywords: Cancer, Diagnosis, Metabolomics, Magnetic Resonance, Nuclear.

---

* **Corresponding author Dinesh Kumar:** Central University of Gujrat, School of Chemical Sciences, Gandhinagar-382030, India, Tel: +919928108023; E-mail: dinesh.kumar@cug.ac.in

INTRODUCTION

Cancer is characterized by the irregular growth of cells within the body. The cancer cells can spread to other organs from the primary site *via* the bloodstream or lymphatic system, causing secondary growth or metastases. Factors known to increase the risk of cancer include the use of tobacco, excessive use of alcohol, exposure to harmful chemicals, radiation, genetic mutation, hormonal imbalance, and a weak immune system. Approximately 8.2 million people die from cancer every year, globally [1 - 6]. Timely diagnosis of cancer is of paramount importance, as is developing techniques that could be effectively used for treating the disease [7].

Metabolomics is an area in analytical biochemistry that is applied to cancer research encompassing the "omics" cascade (genomics, transcriptomics, proteomics, and metabolomics) [8]. Genomics deals with the analysis of the genome to comprehend the single genes. The majority of the functional genomics studies focus on the gene expression (transcriptomics) and comprehensive analysis of protein (proteomics). As most of the studies focus only on transcriptomics and proteomics, very few of them work on the metabolomics. However, changes in the metabolomics are the decisive response of genetic alterations, diseases, and environmental influences. Therefore, quantitative and comprehensive analysis of metabolomics is an attractive tool for diagnosis of diseases [9, 10].

Metabolomics studies the changes in the biological molecules caused by intrinsic and extrinsic factors of various biological pathways. In the recent times, metabolomics has been extensively used for the screening of metabolic biomarkers in various diseases, including cancers of the breast, liver, prostate, lung, pancreas, kidney, and colorectal cancers [11 - 15]. The metabolic profile of cancer patients differs from that of healthy individuals and site, stage, and positioning of the tumors have been found to determine the rate of metabolism. Nonetheless, the characteristic increase in the glycolysis pathways is of complex nature which can be addressed by using metabolomics. Identification of specific biomarkers of the tumor is gaining ground as an alternative for diagnosis and treatment. A simple, inexpensive blood assay that provides metabolomic biomarkers with respect to the sequencing of a genome or proteomic analysis could be used for early detection of cancer. While there is no Food and Drug Administration (FDA) approved metabolomics assay, metabolomics is currently being used by the FDA for biomarker identification [16 - 19].

Presently, for the detection of cancer, biomarkers are not only defined on the basis of location such as colon, breast, and lung, *etc.* but also by their molecular characteristics [20 - 24]. The genetic mutation affecting hormonal receptors and oncogenes such as CA-125 is mostly considered as a useful biomarker for ovarian cancer [25] HER-2 (breast) [26] and K-RAS [27 - 29] in colorectal tumors that plays the main role in the pathogenesis of various cancers. However, in some cases, single biomarkers might be used in more than one clinical applications such as BRCA-1/2 mutations used in prognostic and risk assessment biomarkers, prostate-specific antigen (PSA) used in screening and monitoring of prostate cancer, CA-125 used for the monitoring as well as a diagnostic biomarker. There is significant heterogeneity within the current definitions, represented by the fact that patients given a similar diagnosis respond in different ways to the same therapy. Therefore, heterogeneity of the tumor is being addressed with the context of developing targeted therapeutics. Such designed drugs characterize a newer category of therapeutic agents that offer enhanced tumor specificity, efficiency and relatively fewer side-effects. Metabolomics might be made possible with a shift from the one-size fits all approach to a more altered type of cancer drug by identifying sub-group of cancer patients, also categorizing patients who are likely to experience toxicity or develop resistance. Therefore, developing metabolomics-based tools could widen the uses of drugs which are already available in the market. This can be applied in the drug development study for providing insight into the biological mechanisms of action of the drugs and its effects. In this context, Nuclear Magnetic Resonance (NMR) metabolites have been used to ascertain the effectiveness of both radiation and chemotherapy [30].

NMR spectroscopy provides one of the foremost analytical techniques to investigate metabolites. It also identifies time-dependent concentrations in various samples such as biological fluids, extracts and tissue [31]. However, proton NMR spectra have only a limited range of chemical shifts, with overlapping of signals posing a major problem [32 - 35]. Carbon-13, which has a larger chemical shift dispersion, might benefit from sensitive detectors or from isotopic enrichment, at the cost of an improved complexity of spectra because of the coupling in $^{13}C-^{13}C$. Heteronuclear ($^{1}H \rightarrow ^{13}C$) correlation spectroscopy might offer suitable dispersion. Therefore, ^{13}C NMR is not regularly used in such metabolomic study. However, hyperpolarization by dissolution dynamic nuclear polarization (D-DNP) combined with cross polarization (CP) might enhance its popularity in this area [36, 37].

NMR spectroscopy is a precious tool in the modern biochemical and chemical research, with various applications in both biology and medicine. NMR spectrometer helps researchers to categorize metabolites and the intermediate products of metabolic processes in a biological system based on the magnetic properties of their nuclei. It also helps to identify the quantitative and structural characteristics of organic matter such as lactate produced from cancer cells converting glucose. It does this by focusing on the magnetic properties of the nuclei of atoms. NMR has low sensitivity and high limits of detection for metabolites. The *in-vivo* method usually prefers hyperpolarized NMR to characterize metabolism through tracing metabolites as hyperpolarization NMR can increase sensitivity by 10,000 which is an advantage with samples having a very low concentration. NMR spectroscopy finds wide use in chemistry, where it is used for the quantification of organic compounds, structure analysis, and its interpretation. However, today, NMR spectroscopy is one of the three main methods being used for metabolomics studies, together with gas chromatography and liquid chromatography with mass spectrometry (GC-MS, LC-MS). Recently, NMR-based metabolomics has been used successfully in various fields such as toxicology, gene function, early diagnostic, animal model, human samples and metabolic diseases. Fig. (**1**) shows the schematic illustration of NMR-based metabolomics applications. In this chapter, we have discussed the role of NMR metabolomics in drug development and personalized medicine, as we present what has been reported to date and also some basic aspects of NMR metabolomics. Our aim is to provide metabolomics researchers and cancer researchers with an insight into the advantages of NMR metabolomics.

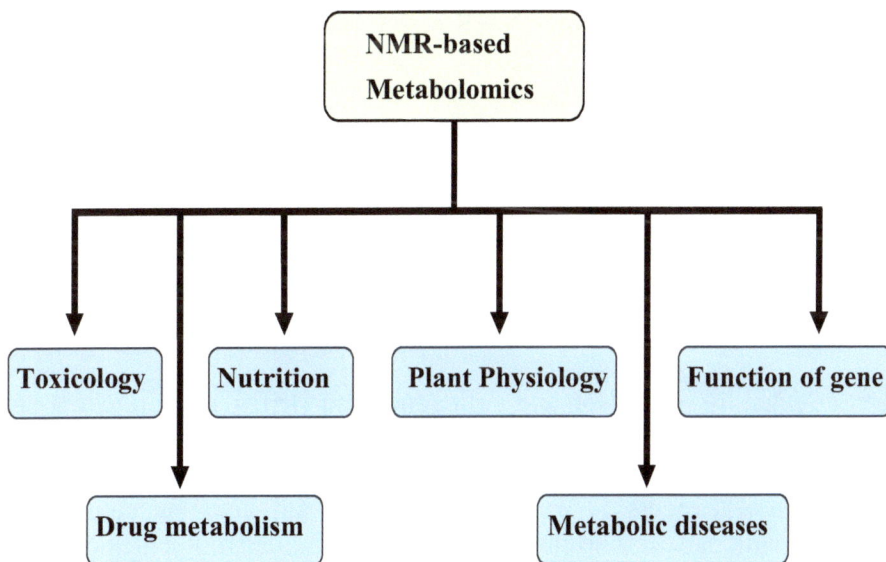

Fig. (1). Schematic illustration of NMR-based metabolomic application.

Basic Principles of NMR Spectroscopy

NMR spectroscopy, commonly recognized as a magnetic resonance spectroscopy (MRS), is a spectroscopic technique that observes magnetic fields around the atomic nuclei. The sample to be analyzed is placed in a magnetic field, and the NMR signal is produced by excitation of the nuclei sample with radiowaves, which is detected by receivers (sensitive radio receivers). The NMR active nuclei such as ^{1}H or ^{13}C absorb electromagnetic radiation at a frequency characteristic of the isotope. The absorption of energy, signal intensity, and resonant frequency is proportional to magnetic field strength. The resonance frequency changes because of the intra-molecular magnetic field around an atom in a molecule, thereby, providing details of the electronic structure and functional groups of the molecules. NMR has replaced conventional wet-chemistry assays such as color reagents and chromatography for their identification of molecules. However, a though recoverable, a relatively large amount of the pure sample around 2-50 mg is required for the analysis. Therefore, NMR spectroscopy is inimitable, tractable, well-resolved, and often extremely predictable for smaller molecules [38, 39].

NMR Spectroscopy in Metabonomics Determination

NMR spectroscopy is an effective tool to investigate tumor metabolism. It is non-destructive, sensitive and a noninvasive, real-time detection technique for biological samples that provide both quantitative and qualitative information of compounds in complex mixtures accurately. The spectral editing technique makes the process effective and flexible. NMR spectroscopy process not only uses ^{1}H spectra but also other nuclei such as ^{13}C, ^{15}N and ^{31}P to provide valuable information about the tumor metabolic pathways. Since these spectroscopy nuclei have wide spectral range, there is no signal overlapping. Since almost all compounds contain proton in the biological samples, ^{1}H-NMR is an extensively used technique to distinguish metabolism between normal and cancerous tissue. As compared with genomics and proteomics, metabolomics can indicate changes in phenotype and function. The quick changes in the metabolomic response to diseases and therapeutic agents might be advantageous. Detection of disease before clinical symptoms manifest facilitate early treatment.

Studies conducted by Nicholson and colleagues and Griffin and Shockcor suggest that NMR and MS metabolomics have the potential ability for rapid identification of cancer cells [32 - 40]. Gowda *et al.* reported metabolomics approaches on different types of cancers [10]. The research group also targeted molecular characteristics with cancer therapeutic agents. The smaller molecules of drugs and monoclonal antibodies dominate the proteasome inhibitors and retinoid [41]. Dunn *et al.* suggested the use of NMR and MS in evaluating the effects of diet and drug in mammals [42]. Several FDA approved inhibitors of growth factor receptor, apoptosis, and signal pathways have been studied by metabolomics [41, 42].

In general, metabolomics is non-invasive or minimally invasive, making it easier to use biofluids. However, the sample size required is larger as compared to mass spectroscopy. While the GC-MS and LC-MS offer relatively higher sensitivity, there are problems associated with the identification of concerned peak, quantitation and also reproducibility.

Current Status of NMR in Cancer Diagnosis

Recently, there have been significant improvements in the instrumentation of NMR such as high-field, ultra-shielded magnets with cryoprobes that provide high sensitivity with smaller magnetic stray fields. The identification is made easier by introducing such modern instruments in the hospitals.

Biofluids, such as plasma, serum, and urine, are most extensively used for analysis. However, use of cerebrospinal fluid, synovial fluid, saliva, and fecal water has also been reported in the clinical practice. The study of intact tissues like biopsy tissue; high-resolution magic angle spinning (HR-MAS) could be used to obtain spectra. The higher resolution is obtained by the spinning of the samples at a particular angle or magic angle. Another advantage of HR-MAS is that there are no extraction procedures, which could affect the metabolic composition. Moreover, changes in the metabolites are difficult to avoid during HR-MAS analysis, as enzymes in the tissue are still active.

The use of a sample from patient-based metabolomics study is likely to be more representative of the processes taking place in the body when compared with cultured cell lines as the cell line is less affected by external factors. However, tumors hardly ever exceed more than 1% of the total body weight, therefore, it is doubtful that overall changes in fluids are only due to cancer. Some changes might be observed due to the contribution of the immune response. Cell lines are preferable for the analysis of newer drugs or drug combinations, as cell lines are less influenced by other factors, as discussed earlier. Therefore, it's decisive the initial step in the drug development and personalized medicine of cancer, thereby, results should be validated in the cancer studies of patients.

Metabonomics is a newer and fast developing area for the identification and quantification of metabolites which are associated with metabolism. The metabolic phenotypes are influenced by factors such as proteomics and genomics, therefore, profiling of metabolic might give accurate information about the individual. The metabolomics could be used as the complementary tool that provides information on the metabolic network, which cannot be obtained from the expression of a gene, genotype, and proteomic analysis. Moreover, metabolomics identifies early signals or biomarkers of cells abnormalities before any changes occur in phenotypes. Therefore, metabolomics has been effectively used for diagnosis of disease.

NMR spectroscopy and mass spectrometry enhance spectral resolution and sensitivity of the metabolomics. NMR spectroscopy is a powerful tool, which is applied in quantification and identification of metabolite. The main advantage of NMR spectroscopy in comparison to other processes is that it is quantitative and there is no sample preparation involved. As protons are present in all metabolites,1H-NMR spectroscopy is an important tool for molecular characterization and has high sensitivity. 1H-NMR spectroscopy is regularly used for monitoring distress in the metabolomics using one-dimensional (1D) spectra followed by data analysis such as partial least-squares discriminant analysis and principal component analysis.

Cancer and Its Types

Cancer is the irregular growth of cells within the body and is the second largest cause of mortality and morbidity, globally. Cancer of the prostate, lung, bronchus, colon, and rectum are prevalent in men, while there was a higher incidence of breast, lung, colon uterine and thyroid cancer in women. Fig. (2) is a schematic representation of normal and cancer cells.

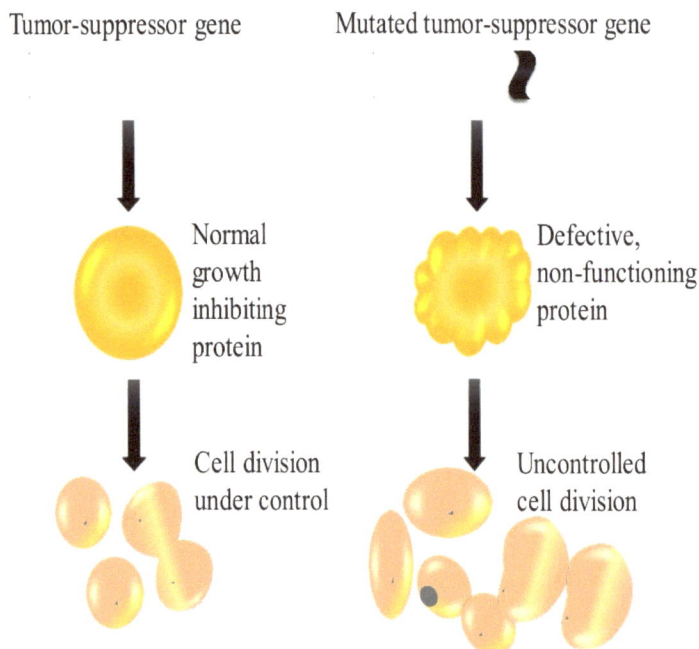

Fig. (2). Schematic illustration of normal and cancer cells differentiation.

Metabolism of cancer cells is different from the normal cells. Moreover, the proliferation of cells and survival is determined by the mechanism of activation of oncogenes or inactivation of tumor suppressor gene. The metabolites are the end products of cellular processes and alteration in metabolic phenotypes is related to changes in the concentration. Identification and quantification of metabolites in plasma by using proton NMR (1H-NMR) is a rapid and noninvasive process, thereby facilitating early detection of cancer.

Breast Cancer

Breast cancer is one of the predominant types of cancer affecting women. According to reports, in Europe, there were 458,337 cases of breast cancer reported in 2012 and around 131,259 were fatal [6]. Statistics indicate that approximately 1.4 million women in the world are suffering from breast cancer, out of those, approximately one-third of patients die. Fig. (**3**) shows schematic illustration of the progression of breast cancer.

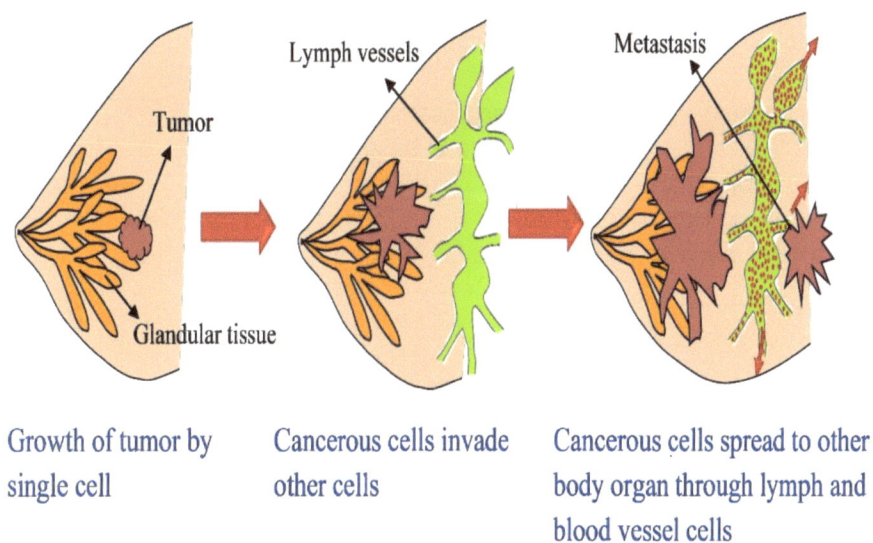

Growth of tumor by single cell

Cancerous cells invade other cells

Cancerous cells spread to other body organ through lymph and blood vessel cells

Fig. (3). Schematic illustration of the progression of breast cancer.

For the diagnosis of breast cancer, many countries introduce mammography which is a rapid diagnostic and has the potential ability for early-stage diagnosis. However, the inconsistency of the results or risk of false positive still remains a concern. In this context, regular screening of women at risk is done to aid early stage detection of cancer [43, 44].

Several studies suggested that detection of low concentration of water-soluble metabolites in breast cancer tissues is possible by using 1H HR-MAS due to the high content of the lipid in tissue. The study compares 1H HR-MAS and *in vitro* NMR data of perchloric extract from the same type of tissue. In 1H HR-MAS spectra, amino acid, lipid, and choline-containing compounds are the major components, and lactate is a minor component, whereas, in PCA extract sample lactose is dominant. The tyrosine and phenylalanine have intense signal in 1H HR-MAS spectra, as opposed to nucleotide signal in the PCA extract [37, 43].

On the other hand, another study suggested that the concentration of the urine metabolite decreases in the breast and ovarian cancer patients. A similar result was also observed in the colon cancer tissues. Excitingly, metabolites such as amino acid mainly increase in the cancer tissue but decrease in the urine sample of cancerous patients. Similar observations were also found in the blood samples, where the concentration of several amino acids decreases compared with that of a healthy person. The changes in the metabolites are mainly hypothetical and are discussed later in the text [44].

Researchers continuously work on the development in sensitivity, the precision of modern explorative technologies and resolution such as metabolites that have the potential ability to identify other risk factors involved in cancer. Moreover, the development is also in the form of prediction modeling at the individual stage of the disease. Biomarker profiling on the basis of Omics is a multifaceted and multi-disciplinary area. The tumor proliferation at the time of diagnosis is possibly one of the dominating factors that affects the survival rates among cancer patients, therefore research on early detection of malignancy is vital.

Lung Cancer

Lung cancer is the most prevalent type of cancer and the five-year survival rates are lower than 15% because the diagnosis is only possible at an advanced stage. Detection of lung cancer at an early stage before metastasizing to the lymph node enhances the five-year survival rate to around 60-80. Thus, there is an urgent need to develop early-stage detection process for lung cancer. Fig. (**4**) schematically illustrates the lung of a healthy person and a cancer patient.

No symptoms were observed during the early phase in a majority of the patients which prevented early detection. Furthermore, various radiological tests, including positron emission tomography (PET), magnetic resonance imaging (MRI), and computed tomography (CT) scan would permit early stage detection. However, these radiological tests are not suitable for screening of the general population because of their high cost. Consequently, there is a need to develop a new process in the detection of lung cancer that contributes to enhancing prognosis. In this context, metabolic markers of cancer in tissue and biofluids have been the focus in the recent year [45 - 48].

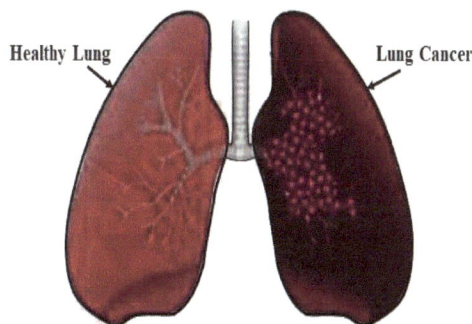

Fig. (4). Schematic illustration of a healthy lung and cancer patient lung.

Blood serum or plasma has been gradually used for the identification of metabolic alterations or metabolic profiling related to different types of cancer. However, very few studies have been reported that focus on plasma or blood serum based metabolic profiling. Maeda *et al.* suggested that the non-small-cell lung cancer (NSCLC) patients were assessed using targeted liquid chromatography with mass spectrometry (LC-MS) on the basis of differences in the plasma amino acid profile compared with that of control (healthy) [49], therefore, LC-MS might be potentially applied for the diagnosis of NSCLC. Another study suggested that abnormal level of the compound lysophosphatidylcholines (lysoPC) isomers with various fatty acyl positions were found in the blood plasma of lung cancer patient. Jordan *et al* suggested that the differentiation of two lung cancer types was possible on the basis of serum metabolomics. On the other hand, metabolic profiling of plasma has been identified for other lung disorders such as pneumonia and sepsis-induced injury that offers distinct metabolic signature related to different pulmonary pathologies.

Among all targeted metabolites, enzyme glycine decarboxylase (GLDC) that degrades the glycine molecules binds with the generated methyl group during purine synthesis. The GLDC expression was increased in lung cancer and decreased in glycine. The GLDC might be used as a biomarker for the lung cancer. On the basis of such discoveries, metabolic signatures might be used in the early detection of cancer by using NMR [45].

Prostate Cancer

Prostate cancer is the fifth leading cause of death associated with cancer, globally and the most diagnosed cancer in men. According to the literature, in approximately 25% of men affected by this disease, bones are the principal target of metastasis. The long latency period of prostate cancer suggests that it is curable. For the cure of prostate cancer, it is crucial to develop an effective and intelligent screening process, for early-stage detection and characterization of cancer. Fig. (**5**) gives a schematic illustration of the prostate of a healthy person and a cancer patient.

Recently, digital rectum analysis and prostate serum antigen (PSA) are frequently used for screening. A serum level of more than 4.0 ng/ml of serum-PSA indicated prostate cancer. However, the value of serum-PSA was found to be below 4ng/ml in some prostate cancer patients. Serum-PSA levels might be affected by various other factors, such as age, urinary tract infection prostatitis, and benign prostate hyperplasia. On the basis of such studies, no cutoff level of PSA could be established that relates to prostate cancer, therefore, these results are not conclusive. Hence, biopsy of the prostate sample is required for the final decision.

On the other hand, the biopsy could give false-negative results in some cases- (1) tumor is small, (2) cancer cells are distributed heterogeneously, (3) and early prostate cancer stage, (4) tumor appears benign [50 - 53]. Therefore, obtained samples during the biopsy of histopathological analysis may not be an envoy of cancer. Therefore, there is a need for newer, specific, sensitive, and economically viable biomarkers to initiate the treatment at an early stage of cancer.

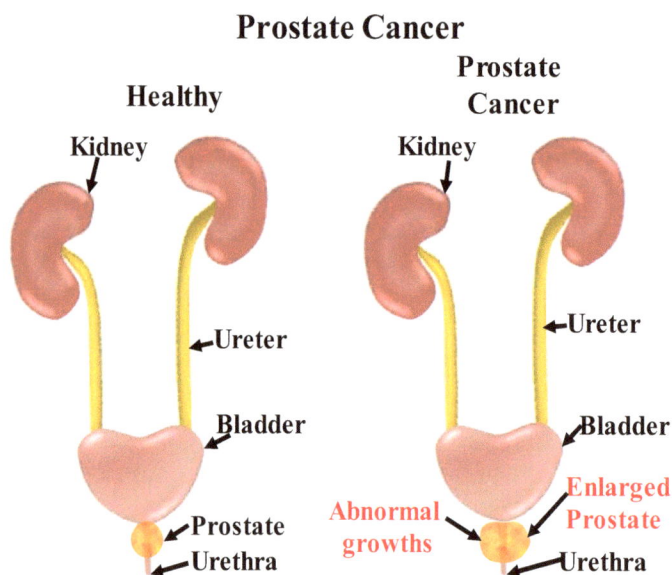

Fig. (5). Schematic illustration of healthy and cancer patient prostate.

Ovarian Cancer

Epithelial ovarian cancer (EOC) is one of the leading causes of mortality and morbidity from gynecologic malignancies, globally, especially, in the United States and Europe. The lower survival rate of EOC patients is because most of the patients are diagnosed at an advanced stage. Therefore, early stage detection is crucial to decrease morbidity and mortality rate. The requirement of such detection techniques or tools is mainly for high-risk EOC women with a family history of cancer or abnormalities in genes mainly BRCA1 and BRCA2 [54 - 58].

Considering the close relation between ovarian pathogenesis and genetic changes, it is apparent that research on the genetic level might offer potential biomarkers. The DNA methylation and histone modification are important in the regulation of gene that plays essential roles in the initiation and progression of the tumor. The DNA methylation of the specific genes in the promoter region can facilitate the prediction of therapeutic response, prognosis and early detection of cancer. The identification of gene-altered by DNA methylation or histone modification in the EOC underneath study such as tumor-specific hypermethylation mainly BRCA1, RAS association domain family protein 1A (RASSF1A), adenomatous polyposis coli (APC), p14ARF, p16INK4a, and death associated protein-kinase (DAPKinase), was found in DNA samples isolated from fifty 50 patients with ovarian tumors. Approximately 82% sensitivity was observed in the hyper methyl; action gene, whereas no hypermethylation or 100% specificity was observed in control samples [25, 54 - 58]. In this context, the various studies suggested that NMR spectroscopy based metabolomics might be used for early detection of EOC patients. Fig. (**6**) is the schematic illustration of the normal and cancerous ovaries.

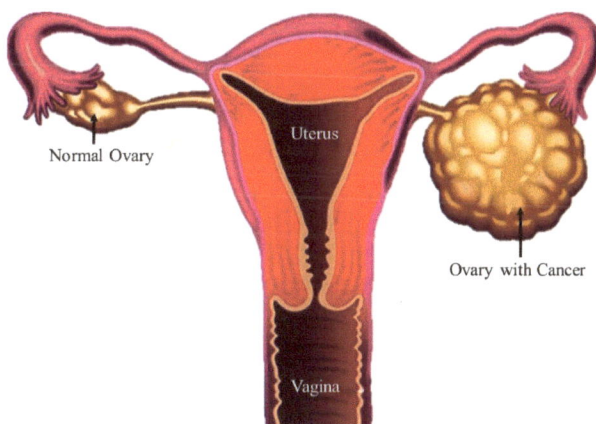

Fig. (6). Schematic illustration of the normal and cancerous ovaries.

Gastric Cancer

Gastric adenocarcinoma (GC) is the fifth most prevalent cancer globally. Estimated around one million individuals are affected by this disease every year, globally. The 5-year survival rates of stage IA, IB tumors are 71% and 57%, respectively [6]. The GC cells convert glucose into lactate in the presence of oxygen [59 - 61]. The citric acid is related to apoptotic pathways in GC cells. These nucleic acids over-expressed in GC attributed active replication. Therefore, urinary profiling of metabolomics offers a non-invasive, economical-viable, effective, and sensitive tool for early and accurate diagnosis of GC cancer [62 - 64]. Fig. (7) shows the schematic illustration of gastric cancer.

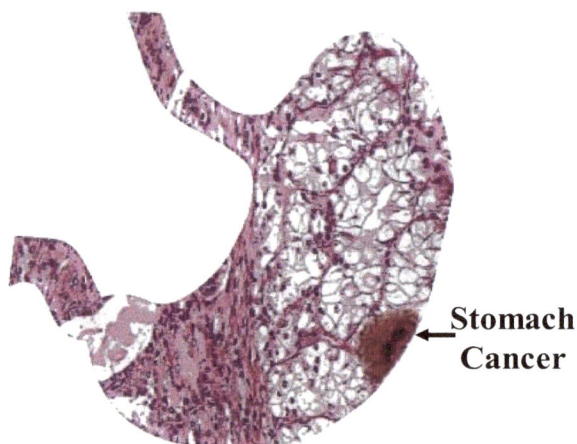

Fig. (7). Schematic illustration of gastric cancer.

In general, almost all types of cancer have low survival rate and the survival rate might be increased by developing newer diagnostic tools that are sensitive, specific, and economically-viable for early-stage detection of various types of cancer.

Growth-Associated Changes in the Energy Metabolism of the Cancer Cell

Heterogeneous cancer and stromal cells incorporated into tumor cells lead to the progression of the disease. The heterogeneity of the tumor develops because of the epigenetic and genetic alteration within the cancer cells.

New biochemical and molecular biological tools and studies in cancer cell metabolism have expanded our understanding of the mechanisms and functional consequences of tumor-associated metabolic alterations. Some biosynthetic and bioenergetics molecules require such as a high ATP requirement, lipids, amino acids, and reducing equivalents such as NADH, NADPH, FADH2, for maintaining the cell proliferation. The process of energy producing and consuming is balanced by the cancer cell, therefore significant changes occur in nutrient and oxygen growth during the progression of the tumor. The growth of the tumor and the progression of the disease lead to metastatic required metabolic interaction with the various types of cells that include primary and metastatic tumor micro-environments. Every stage of disease progression contributes to a high degree of metabolic heterogeneity. The metabolic adaptation and flexibility demonstrated by cancerous cells are critical for their ability to grow in discrete metastatic microenvironments [65 - 70]. Fig. (**8**) shows growth of the tumor and the progression of the disease which need complex metabolic interactions with several types of cells.

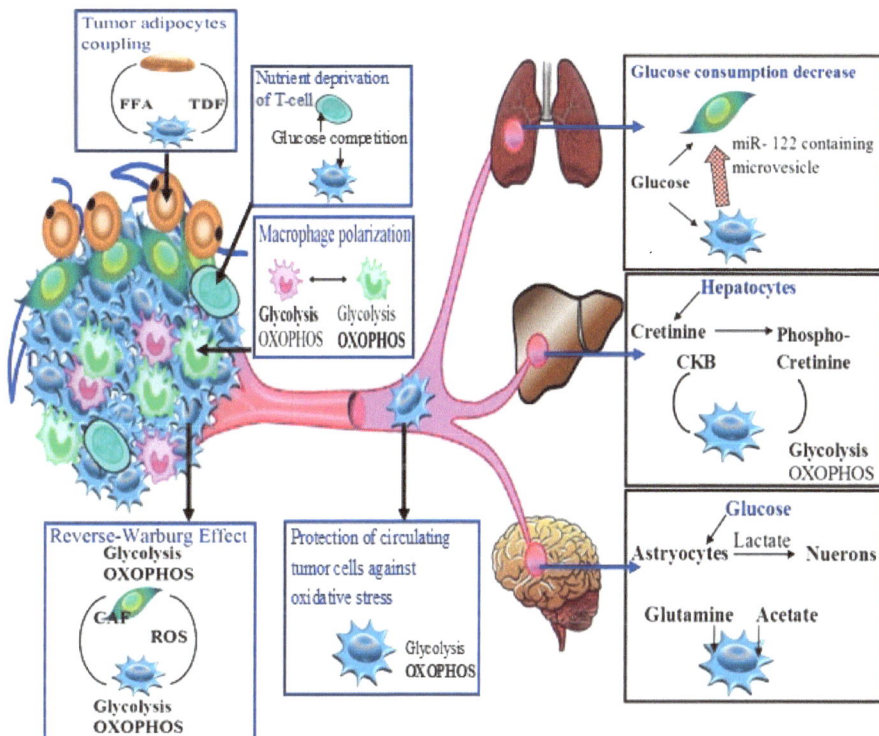

Fig. (8). Growth of tumor and progression of disease needs complex metabolic interactions with several types of cells, reactive oxygen species (ROS), tumor-derived factors (TDF), free fatty acids (FFA), creatinine kinase brain-type (CKB).

Cancerous cells can exhibit distinctive metabolic features that depend on the origin of the tissue. Undeniably, liver, lung, colorectal cancers, and leukemias depend on the glycolysis, while, lymphomas, melanomas, and glioblastomas are distinguished as oxidative tumors. Interestingly, evidence supports that cancer cells-type and subtypes might be also adopting the different metabolic approach, for example, triple-negative breast cancers (TNBC) that demonstrate a Warburg-like phenotype, while estrogen receptor-positive breast cancers might depend on Oxidative phosphorylation. The higher glycolytic rates associated with enhanced TNBC proliferation index activate oxidative metabolism in TNBC that reduces both growthof tumor and metastasis. The analysis of the metabolic profile of several breast cancer shows different metastatic abilities.

Otto Warburg *et al* (1930) observed the changes in the metabolism within the cancer cells. Otto Warburg *et al* (1956) suggested that in tumors and other proliferating or developing cells, the rate of glucose uptake dramatically increases and lactate is produced even in the presence of oxygen. The research group of Warburg also observed that the consumption of oxygen is approximately the same in the cancerous cells in comparison with that of normal cells. Cancerous cells prefer aerobic glycolysis than the oxidative phosphorylation even at normal oxygen condition. This research group understood that respiration was impaired because of mitochondrial defects. However, various research groups reported that for the large production of lactic acid cancerous cells did not use oxidative phosphorylation.

The hypothesis on the adaptation of fermentative glucose metabolism by cancerous cells has numerous advantages. The adaptation of proliferated cells required more uptakes of nutrients such as amino acid for the synthesis of protein; duplication of DNA required nucleic acid, and lipids for bio-membrane synthesis. Cancer cells adopt glycolysis to accelerate growth. The acidic environment provided by glycolysis is harmful to normal cells, but has no effect on tumor cells. On the basis of the above discussion, we understand the importance of glycolysis in the defense mechanism for the cancerous cells. The above hypothesis based on the mathematical models and empirical observations is supported by the reports on the lymphocytes activating glycolysis during the fast-growing condition.

The process of glycolysis produces less reactive oxygen species (ROS), therefore, the genome of cancer cells might be evaded from damaging of cells. The less production of ROS would result in developing apoptosis resistance in tumors. The plausible defense mechanism in cancerous cells has been established and cancer cells expand survival advantages concurrently. Glycolysis might be generating ATP faster than oxidative phosphorylation. Nonetheless, it remains a challenge to know the minimum ATP molecules required for cell proliferation [71 - 76].

^{31}P-NMR Spectra for Prediction and Detection of Therapeutic Response

Localized high-resolution ^{31}P-NMR spectra can be used to study metabolic changes in animals and humans. The spectra deal with the cellular phosphate compounds which participate in tissue metabolism. Only those changes that can provide some useful information about the therapeutic response such as perfusion, energetics, and pH, are measurable.

^{31}P-NMR spectra are used for the determination of phosphate compounds such as energy metabolism (NTP, PCr, Pi), intracellular pH and phospholipid metabolism.

It is preferably matched to *in-vivo* study tumor energy metabolism. Notably, the pH of the tumor can be determined from the chemical shift data using Pi peak. However, most clinical studies have declined its use as poor sensitivity is the major drawback of ^{31}P NMR. Usually, the cancer cell has a relatively lower bioenergetics state while high-energy phosphates production such as NTP and PCr mainly depends on accessible glucose and oxygen content delivered to the tumor by blood vessels. Therefore, energy metabolism is tightly coupled with tumor blood flow and decreases in hypoxic regions. These remarks raise serious doubts about the potential diagnostic utility of *in vivo*^{31}P NMR spectroscopy of tumors with spectral characteristics. ^{31}P NMR spectra are useful for the interpretation of later stage of tumor in which spectral characteristics can be easily distinguished. However, the ^{31}P NMR spectrum of the tumor may be a very sensitive noninvasive indicator of the state of oxygenation of the tumor [77 - 79].

It is only at later stages that the ^{31}P NMR spectral characteristics of the tumor become readily distinguishable from those of normal tissues, but by then the diagnosis can probably be more readily accomplished by other means.

FUTURE PROSPECTS AND CONCLUSION

This book chapter highlights the diversity of metabolic approaches adopted by cancerous cells. The metabolic profiling might have a relation between different cancerous cells population within the tumor. Moreover, aerobic glycolysis, metastatic tumor cells might cause various metabolic changes that allow cancer cells to respond rapidly upon changes in the metabolic demands. In this book chapter, we discuss many metabolic approaches which are used by cancerous cells for growth of tumor or progression of the disease. Metabolic reprogramming of cancer cells is accepted worldwide. However, researchers face many challenges as to how reprogramming is related to the progression of cancer. The alteration in metabolism during metastasis process is also not established.

NMR-spectroscopic based metabolomics have potential tools of system biology that are applied in many aspects of biological research such as clinical diagnostic and prognosis of various types of cancer and also other diseases. The combination of different platforms mainly NMR and MS-based metabolomics and interpretation of various data such as proteomics, genomics, and transcriptomics, provide a strategic approach to metabolic process that contributes to health and disease. The NMR and MS-based metabolomic might be beneficial in the diagnosis and treatment of cancer treatment at an early stage of the disease.

CONSENT FOR PUBLICATION

Not applicable.

CONFLICT OF INTEREST

The author(s) declare no competing financial interests.

ACKNOWLEDGEMENTS

We gratefully acknowledge support from the Ministry of Science and Technology and Department of Science and Technology, Government of India under the Scheme of Establishment of Women Technology Park, for providing the necessary financial support to carry out this study vide letter No, F. No SEED/WTP/063/2014

REFERENCES

[1]   Kang, Y.P.; Ward, N.P.; DeNicola, G.M. Recent advances in cancer metabolism: a technological perspective. *Exp. Mol. Med.,* **2018**, *50*(4), 31.
[http://dx.doi.org/10.1038/s12276-018-0027-z] [PMID: 29657324]

[2]   Siegel, R.; Ward, E.; Brawley, O.; Jemal, A. Cancer statistics, 2011: the impact of eliminating socioeconomic and racial disparities on premature cancer deaths. *CA Cancer J. Clin.,* **2011**, *61*(4), 212-236.
[http://dx.doi.org/10.3322/caac.20121] [PMID: 21685461]

[3]   Torre, L.A.; Bray, F.; Siegel, R.L.; Ferlay, J.; Lortet-Tieulent, J.; Jemal, A. Global cancer statistics, 2012. *CA Cancer J. Clin.,* **2015**, *65*(2), 87-108.
[http://dx.doi.org/10.3322/caac.21262] [PMID: 25651787]

[4]   Chan, A.W.; Mercier, P.; Schiller, D.; Bailey, R.; Robbins, S.; Eurich, D.T.; Sawyer, M.B.; Broadhurst, D. (1)H-NMR urinary metabolomic profiling for diagnosis of gastric cancer. *Br. J. Cancer,* **2016**, *114*(1), 59-62.
[http://dx.doi.org/10.1038/bjc.2015.414] [PMID: 26645240]

[5]   Blagg, J.; Workman, P. Chemical biology approaches to target validation in cancer. *Curr. Opin. Pharmacol.,* **2014**, *17*, 87-100.
[http://dx.doi.org/10.1016/j.coph.2014.07.007] [PMID: 25175311]

[6]   Ferlay, J.; Soerjomataram, I.; Dikshit, R.; Eser, S.; Mathers, C.; Rebelo, M.; Parkin, D.M.; Forman, D.; Bray, F. Cancer incidence and mortality worldwide: sources, methods and major patterns in GLOBOCAN 2012. *Int. J. Cancer,* **2015**, *136*(5), E359-E386.
[http://dx.doi.org/10.1002/ijc.29210] [PMID: 25220842]

[7]   Alberts, B.; Johnson, A.; Lewis, J. *Molecular Biology of the Cell,* (4th.), New York: Garland Science; 2002. Cancer Treatment: Present and Future https://www.ncbi.nlm.nih.gov/books/NBK26811/

[8]   Rai, V.; Mukherjee, R.; Ghosh, A.K.; Routray, A.; Chakraborty, C. "Omics" in oral cancer: New approaches for biomarker discovery. *Arch. Oral Biol.,* **2018**, *87*, 15-34.
[http://dx.doi.org/10.1016/j.archoralbio.2017.12.003] [PMID: 29247855]

[9]   Li, T.; Deng, P. Nuclear Magnetic Resonance technique in tumor metabolism. *Genes Dis.,* **2017**, *4*, 28-36.
[http://dx.doi.org/10.1016/j.gendis.2016.12.001]

[10]   Gowda, G.A.N.; Zhang, S.; Gu, H.; Asiago, V.; Shanaiah, N.; Raftery, D. Metabolomics-based methods for early disease diagnostics. *Expert Rev. Mol. Diagn.,* **2008**, *8*(5), 617-633.
[http://dx.doi.org/10.1586/14737159.8.5.617] [PMID: 18785810]

[11]   Aboud, O.A.; Weiss, R.H. New opportunities from the cancer metabolome. *Clin. Chem.,* **2013**, *59*(1), 138-146.

[http://dx.doi.org/10.1373/clinchem.2012.184598] [PMID: 23150057]

[12]   Nicholson, J.K.; Lindon, J.C.; Holmes, E. 'Metabonomics': understanding the metabolic responses of living systems to pathophysiological stimuli *via* multivariate statistical analysis of biological NMR spectroscopic data. *Xenobiotica,* **1999**, *29*(11), 1181-1189.
[http://dx.doi.org/10.1080/004982599238047] [PMID: 10598751]

[13]   Spratlin, J.L.; Serkova, N.J.; Eckhardt, S.G. Clinical applications of metabolomics in oncology: a review. *Clin. Cancer Res.,* **2009**, *15*(2), 431-440.
[http://dx.doi.org/10.1158/1078-0432.CCR-08-1059] [PMID: 19147747]

[14]   Soga, T. Cancer metabolism: key players in metabolic reprogramming. *Cancer Sci.,* **2013**, *104*(3), 275-281.
[http://dx.doi.org/10.1111/cas.12085] [PMID: 23279446]

[15]   Bingol, K.; Brüschweiler, R. Knowns and unknowns in metabolomics identified by multidimensional NMR and hybrid MS/NMR methods. *Curr. Opin. Biotechnol.,* **2017**, *43*, 17-24.
[http://dx.doi.org/10.1016/j.copbio.2016.07.006] [PMID: 27552705]

[16]   Halama, A. Metabolomics in cell culture a strategy to study crucial metabolic pathways in cancer development and the response to treatment. *Arch. Biochem. Biophys.,* **2014**, *564*, 100-109.
[http://dx.doi.org/10.1016/j.abb.2014.09.002] [PMID: 25218088]

[17]   Ramautar, R.; Berger, R.; van der Greef, J.; Hankemeier, T. Human metabolomics: strategies to understand biology. *Curr. Opin. Chem. Biol.,* **2013**, *17*(5), 841-846.
[http://dx.doi.org/10.1016/j.cbpa.2013.06.015] [PMID: 23849548]

[18]   Hart, C.D.; Tenori, L.; Luchinat, C.; Di Leo, A. Metabolomics in breast cancer: current status and perspectives. *Adv. Exp. Med. Biol.,* **2016**, *882*, 217-234.
[http://dx.doi.org/10.1007/978-3-319-22909-6_9] [PMID: 26987537]

[19]   McCartney, A.; Vignoli, A.; Biganzoli, L.; Love, R.; Tenori, L.; Luchinat, C.; Di Leo, A. Metabolomics in breast cancer: A decade in review. *Cancer Treat. Rev.,* **2018**, *67*, 88-96.
[http://dx.doi.org/10.1016/j.ctrv.2018.04.012] [PMID: 29775779]

[20]   Maruvada, P.; Wang, W.; Wagner, P.D.; Srivastava, S. Biomarkers in molecular medicine: cancer detection and diagnosis. *Biotechniques,* **2005** Suppl., 9-15.
[http://dx.doi.org/10.2144/05384SU04] [PMID: 16528918]

[21]   Lima, A.R.; Bastos, Mde.L.; Carvalho, M.; Guedes de Pinho, P. Biomarker discovery in human prostate cancer: an update in metabolomics studies. *Transl. Oncol.,* **2016**, *9*(4), 357-370.
[http://dx.doi.org/10.1016/j.tranon.2016.05.004] [PMID: 27567960]

[22]   Monteiro, M.; Carvalho, M.; de Lourdes Bastos, M.; de Pinho, P. Biomarkers in renal cell carcinoma: a metabolomics approach. *Metabolomics,* **2014**, *10*, 1210-1222.
[http://dx.doi.org/10.1007/s11306-014-0659-5]

[23]   Pan, Z; Gu, H; Raftery, D. *NMR-Based Metabolomics Technology in Biomarker Research.,*
[http://dx.doi.org/10.1002/9780470048672.wecb637]

[24]   Zhang, A.H.; Sun, H.; Qiu, S.; Wang, X.J. NMR-based metabolomics coupled with pattern recognition methods in biomarker discovery and disease diagnosis. *Magn. Reson. Chem.,* **2013**, *51*(9), 549-556.
[http://dx.doi.org/10.1002/mrc.3985] [PMID: 23828598]

[25]   Zhang, B.; Cai, F.F.; Zhong, X.Y. An overview of biomarkers for the ovarian cancer diagnosis. *Eur. J. Obstet. Gynecol. Reprod. Biol.,* **2011**, *158*(2), 119-123.
[http://dx.doi.org/10.1016/j.ejogrb.2011.04.023] [PMID: 21632171]

[26]   Duffy, M.J.; Harbeck, N.; Nap, M.; Molina, R.; Nicolini, A.; Senkus, E.; Cardoso, F. Clinical use of biomarkers in breast cancer: Updated guidelines from the European Group on Tumor Markers (EGTM). *Eur. J. Cancer,* **2017**, *75*, 284-298.
[http://dx.doi.org/10.1016/j.ejca.2017.01.017] [PMID: 28259011]

[27]   Zarkavelis, G.; Boussios, S.; Papadaki, A.; Katsanos, K.H.; Christodoulou, D.K.; Pentheroudakis, G. Current and future biomarkers in colorectal cancer. *Ann. Gastroenterol.,* **2017**, *30*(6), 613-621. [PMID: 29118555]

[28]   Lech, G.; Słotwiński, R.; Słodkowski, M.; Krasnodębski, I.W. Colorectal cancer tumour markers and biomarkers: Recent therapeutic advances. *World J. Gastroenterol.,* **2016**, *22*(5), 1745-1755. [http://dx.doi.org/10.3748/wjg.v22.i5.1745] [PMID: 26855534]

[29]   Das, V.; Kalita, J.; Pal, M. Predictive and prognostic biomarkers in colorectal cancer: A systematic review of recent advances and challenges. *Biomed. Pharmacother.,* **2017**, *87*, 8-19. [http://dx.doi.org/10.1016/j.biopha.2016.12.064] [PMID: 28040600]

[30]   Palmnas, M.S.A.; Vogel, H.J. The future of NMR metabolomics in cancer therapy: towards personalizing treatment and developing targeted drugs? *Metabolites,* **2013**, *3*(2), 373-396. [http://dx.doi.org/10.3390/metabo3020373] [PMID: 24957997]

[31]   Serkova, N.J.; Niemann, C.U. Pattern recognition and biomarker validation using quantitative 1H-NMR-based metabolomics. *Expert Rev. Mol. Diagn.,* **2006**, *6*(5), 717-731. [http://dx.doi.org/10.1586/14737159.6.5.717] [PMID: 17009906]

[32]   Nicholson, J.K.; Wilson, I.D. High resolution proton magnetic resonance spectroscopy of biological fluids. *Prog. Nucl. Magn. Reson. Spectrosc.,* **1989**, *21*(4), 449-501. [http://dx.doi.org/10.1016/0079-6565(89)80008-1]

[33]   Gebregiworgis, T.; Powers, R. Application of NMR metabolomics to search for human disease biomarkers. *Comb. Chem. High Throughput Screen.,* **2012**, *15*(8), 595-610. [http://dx.doi.org/10.2174/138620712802650522] [PMID: 22480238]

[34]   Yang, C.; Richardson, A.D.; Osterman, A.; Smith, J.W. Profiling of central metabolism in human cancer cells by two-dimensional NMR, GC-MS analysis, and isotopomer modeling. *Metabolomics,* **2008**, *4*(1), 13-29. [http://dx.doi.org/10.1007/s11306-007-0094-y]

[35]   de Graaf, R.A.; Behar, K.L. Quantitative 1H NMR spectroscopy of blood plasma metabolites. *Anal. Chem.,* **2003**, *75*(9), 2100-2104. [http://dx.doi.org/10.1021/ac020782+] [PMID: 12720347]

[36]   Curzon, E.H.; Hawkes, G.E.; Randall, E.W.; Britton, H.G.; Fazakerley, G.V. Conformational analysis of 2- and 3-phosphoglyceric acids by 1H and 13C nuclear magnetic resonance spectroscopy. *J. Chem. Soc. Perkin Trans.,* **1981**, *3*, 494-499. [http://dx.doi.org/10.1039/p29810000494]

[37]   Harris, T.; Degani, H.; Frydman, L. Hyperpolarized 13C NMR studies of glucose metabolism in living breast cancer cell cultures. *NMR Biomed.,* **2013**, *26*(12), 1831-1843. [http://dx.doi.org/10.1002/nbm.3024] [PMID: 24115045]

[38]   Marion, D. An introduction to biological NMR spectroscopy. *Mol. Cell. Proteomics,* **2013**, *12*(11), 3006-3025. [http://dx.doi.org/10.1074/mcp.O113.030239] [PMID: 23831612]

[39]   Gerothanassis, I.P.; Troganis, A.; Exarchou, V.; Barbarossou, K. nuclear magnetic resonance (nmr) spectroscopy: basic principles and phenomena, and their applications to chemistry, biology and medicine. *Chem. Educ. Res. Pract.,* **2002**, *3*(2), 229-252. [http://dx.doi.org/10.1039/B2RP90018A]

[40]   Griffin, J.L.; Atherton, H.; Shockcor, J.; Atzori, L. Metabolomics as a tool for cardiac research. *Nat. Rev. Cardiol.,* **2011**, *8*(11), 630-643. [http://dx.doi.org/10.1038/nrcardio.2011.138] [PMID: 21931361]

[41]   Wei, S.; Liu, L.; Zhang, J.; Bowers, J.; Gowda, G.A.; Seeger, H.; Fehm, T.; Neubauer, H.J.; Vogel, U.; Clare, S.E.; Raftery, D. Metabolomics approach for predicting response to neoadjuvant chemotherapy for breast cancer. *Mol. Oncol.,* **2013**, *7*(3), 297-307.

[http://dx.doi.org/10.1016/j.molonc.2012.10.003] [PMID: 23142658]

[42]   Dunn, W.B.; Broadhurst, D.I.; Atherton, H.J.; Goodacre, R.; Griffin, J.L. Systems level studies of mammalian metabolomes: the roles of mass spectrometry and nuclear magnetic resonance spectroscopy. *Chem. Soc. Rev.,* **2011**, *40*(1), 387-426.
        [http://dx.doi.org/10.1039/B906712B] [PMID: 20717559]

[43]   Whitehead, T.L.; Kieber-Emmons, T. Applying *in vitro* NMR spectroscopy and H-1 NMR metabonomics to breast cancer characterization and detection. *Prog. Nucl. Magn. Reson. Spectrosc.,* **2005**, *47*, 165-174.
        [http://dx.doi.org/10.1016/j.pnmrs.2005.09.001]

[44]   Slupsky, C.M.; Steed, H.; Wells, T.H.; Dabbs, K.; Schepansky, A.; Capstick, V.; Faught, W.; Sawyer, M.B. Urine metabolite analysis offers potential early diagnosis of ovarian and breast cancers. *Clin. Cancer Res.,* **2010**, *16*(23), 5835-5841.
        [http://dx.doi.org/10.1158/1078-0432.CCR-10-1434] [PMID: 20956617]

[45]   K, OShea; SJS, Cameron; KE, Lewis; C, Lu; LAJ, Mur Metabolomic-based biomarker discovery for non-invasive lung cancer screening: A case study. *BBA-GEN. SUBJECTS,* **2016**, *11*(1860), 2682-2887.

[46]   Carrola, J.; Rocha, C.M.; Barros, A.S.; Gil, A.M.; Goodfellow, B.J.; Carreira, I.M.; Bernardo, J.; Gomes, A.; Sousa, V.; Carvalho, L.; Duarte, I.F. Metabolic signatures of lung cancer in biofluids: NMR-based metabonomics of urine. *J. Proteome Res.,* **2011**, *10*(1), 221-230.
        [http://dx.doi.org/10.1021/pr100899x] [PMID: 21058631]

[47]   Rocha, C.M.; Barros, A.S.; Goodfellow, B.J.; Carreira, I.M.; Gomes, A.; Sousa, V.; Bernardo, J.; Carvalho, L.; Gil, A.M.; Duarte, I.F. NMR metabolomics of human lung tumours reveals distinct metabolic signatures for adenocarcinoma and squamous cell carcinoma. *Carcinogenesis,* **2015**, *36*(1), 68-75.
        [http://dx.doi.org/10.1093/carcin/bgu226] [PMID: 25368033]

[48]   Goja, A; Geamanu, A; Sadaat, N; Gupta, S. Potential 1H NMR urinary metabolomic biomarkers for Lung cancer in hamster model **2011**, *1*(25), 984-984.

[49]   Maeda, R.; Yoshida, J.; Ishii, G.; Hishida, T.; Nishimura, M.; Nagai, K. Prognostic impact of intratumoral vascular invasion in non-small cell lung cancer patients. *Thorax,* **2010**, *65*(12), 1092-1098.
        [http://dx.doi.org/10.1136/thx.2010.141861] [PMID: 20971984]

[50]   Stabler, S.; Koyama, T.; Zhao, Z.; Martinez-Ferrer, M.; Allen, R.H.; Luka, Z.; Loukachevitch, L.V.; Clark, P.E.; Wagner, C.; Bhowmick, N.A. Serum methionine metabolites are risk factors for metastatic prostate cancer progression. *PLoS One,* **2011**, *6*(8), e22486.
        [http://dx.doi.org/10.1371/journal.pone.0022486] [PMID: 21853037]

[51]   Kumar, D.; Gupta, A.; Nath, K. NMR-based metabolomics of prostate cancer: a protagonist in clinical diagnostics. Expert review of molecular diagnostics. *Biochimica Et Biophysica,* **2016**, *1860*, 2682-2687.

[52]   Trock, B.J. Application of metabolomics to prostate cancer. *Urol. Oncol.,* **2011**, *29*(5), 572-581.
        [http://dx.doi.org/10.1016/j.urolonc.2011.08.002] [PMID: 21930089]

[53]   DeFeo, E.M.; Wu, C.L.; McDougal, W.S.; Cheng, L.L. A decade in prostate cancer: from NMR to metabolomics. *Nat. Rev. Urol.,* **2011**, *8*(6), 301-311.
        [http://dx.doi.org/10.1038/nrurol.2011.53] [PMID: 21587223]

[54]   Ueland, F.R. A Perspective on Ovarian Cancer Biomarkers: Past, Present and Yet-To-Come. *Diagnostics (Basel),* **2017**, *7*(1), 14.
        [http://dx.doi.org/10.3390/diagnostics7010014] [PMID: 28282875]

[55]   Odunsi, K.; Wollman, R.M.; Ambrosone, C.B.; Hutson, A.; McCann, S.E.; Tammela, J.; Geisler, J.P.; Miller, G.; Sellers, T.; Cliby, W.; Qian, F.; Keitz, B.; Intengan, M.; Lele, S.; Alderfer, J.L. Detection

of epithelial ovarian cancer using 1H-NMR-based metabonomics. *Int. J. Cancer,* **2005**, *113*(5), 782-788.
[http://dx.doi.org/10.1002/ijc.20651] [PMID: 15499633]

[56]   Garcia, E.; Andrews, C.; Hua, J.; Kim, H.L.; Sukumaran, D.K.; Szyperski, T.; Odunsi, K. Diagnosis of early stage ovarian cancer by 1H NMR metabonomics of serum explored by use of a microflow NMR probe. *J. Proteome Res.,* **2011**, *10*(4), 1765-1771.
[http://dx.doi.org/10.1021/pr101050d] [PMID: 21218854]

[57]   Belki, D.; Belki, K. *In vivo* magnetic resonance spectroscopy for ovarian cancer diagnostics: quantification by the fast Padé transform. *J. Math. Chem.,* **2017**, *55*, 349-405.
[http://dx.doi.org/10.1007/s10910-016-0694-8]

[58]   Garcia, E.; Andrews, C.; Hua, J.; Kim, H.L.; Sukumaran, D.K.; Szyperski, T.; Odunsi, K. Diagnosis of early stage ovarian cancer by 1H NMR metabonomics of serum explored by use of a microflow NMR probe. *J. Proteome Res.,* **2011**, *10*(4), 1765-1771.
[http://dx.doi.org/10.1021/pr101050d] [PMID: 21218854]

[59]   Hirayama, A.; Kami, K.; Sugimoto, M.; Sugawara, M.; Toki, N.; Onozuka, H.; Kinoshita, T.; Saito, N.; Ochiai, A.; Tomita, M.; Esumi, H.; Soga, T. Quantitative metabolome profiling of colon and stomach cancer microenvironment by capillary electrophoresis time-of-flight mass spectrometry. *Cancer Res.,* **2009**, *69*(11), 4918-4925.
[http://dx.doi.org/10.1158/0008-5472.CAN-08-4806] [PMID: 19458066]

[60]   Cai, Z.; Zhao, J.S.; Li, J.J.; Peng, D.N.; Wang, X.Y.; Chen, T.L.; Qiu, Y.P.; Chen, P.P.; Li, W.J.; Xu, L.Y.; Li, E.M.; Tam, J.P.M.; Qi, R.Z.; Jia, W.; Xie, D. A combined proteomics and metabolomics profiling of gastric cardia cancer reveals characteristic dysregulations in glucose metabolism. *Mol. Cell. Proteomics,* **2010**, *9*(12), 2617-2628.
[http://dx.doi.org/10.1074/mcp.M110.000661] [PMID: 20699381]

[61]   Hu, K.; Chen, F. Identification of significant pathways in gastric cancer based on protein-protein interaction networks and cluster analysis. *Genet. Mol. Biol.,* **2012**, *35*(3), 701-708.
[http://dx.doi.org/10.1590/S1415-47572012005000045] [PMID: 23055812]

[62]   Chan, A.W.; Mercier, P.; Schiller, D.; Bailey, R.; Robbins, S.; Eurich, D.T.; Sawyer, M.B.; Broadhurst, D. (1)H-NMR urinary metabolomic profiling for diagnosis of gastric cancer. *Br. J. Cancer,* **2016**, *114*(1), 59-62.
[http://dx.doi.org/10.1038/bjc.2015.414] [PMID: 26645240]

[63]   Abbassi-Ghadi, N.; Kumar, S.; Huang, J.; Goldin, R.; Takats, Z.; Hanna, G.B. Metabolomic profiling of oesophago-gastric cancer: a systematic review. *Eur. J. Cancer,* **2013**, *49*(17), 3625-3637.
[http://dx.doi.org/10.1016/j.ejca.2013.07.004] [PMID: 23896378]

[64]   Ramachandran, G.K.; Yong, W.P.; Yeow, C.H. Identification of Gastric Cancer Biomarkers Using 1H Nuclear Magnetic Resonance Spectrometry. *PLoS One,* **2016**, *11*(9), e0162222.
[http://dx.doi.org/10.1371/journal.pone.0162222] [PMID: 27611679]

[65]   Obre, E.; Rossignol, R. Emerging concepts in bioenergetics and cancer research: metabolic flexibility, coupling, symbiosis, switch, oxidative tumors, metabolic remodeling, signaling and bioenergetic therapy. *Int. J. Biochem. Cell Biol.,* **2015**, *59*, 167-181.
[http://dx.doi.org/10.1016/j.biocel.2014.12.008] [PMID: 25542180]

[66]   Elia, I.; Schmieder, R.; Christen, S.; Fendt, S.M. Organ-specific cancer metabolism and its potential for therapy. *Handb. Exp. Pharmacol.,* **2016**, *233*, 321-353.
[http://dx.doi.org/10.1007/164_2015_10] [PMID: 25912014]

[67]   Ma, R.; Zhang, W.; Tang, K.; Zhang, H.; Zhang, Y.; Li, D.; Li, Y.; Xu, P.; Luo, S.; Cai, W.; Ji, T.; Katirai, F.; Ye, D.; Huang, B. Switch of glycolysis to gluconeogenesis by dexamethasone for treatment of hepatocarcinoma. *Nat. Commun.,* **2013**, *4*(2508), 2508.
[http://dx.doi.org/10.1038/ncomms3508] [PMID: 24149070]

[68]   Amann, T.; Hellerbrand, C. GLUT1 as a therapeutic target in hepatocellular carcinoma. *Expert Opin.*

Ther. Targets, **2009**, *13*(12), 1411-1427.
[http://dx.doi.org/10.1517/14728220903307509] [PMID: 19874261]

[69]   Long, J.; Lang, Z.W.; Wang, H.G.; Wang, T.L.; Wang, B.E.; Liu, S.Q. Glutamine synthetase as an early marker for hepatocellular carcinoma based on proteomic analysis of resected small hepatocellular carcinomas. *HBPD INT,* **2010**, *9*(3), 296-305.
[PMID: 20525558]

[70]   Calvisi, D.F.; Wang, C.; Ho, C.; Ladu, S.; Lee, S.A.; Mattu, S.; Destefanis, G.; Delogu, S.; Zimmermann, A.; Ericsson, J.; Brozzetti, S.; Staniscia, T.; Chen, X.; Dombrowski, F.; Evert, M. Increased lipogenesis, induced by AKT-mTORC1-RPS6 signaling, promotes development of human hepatocellular carcinoma. *Gastroenterology,* **2011**, *140*(3), 1071-1083.
[http://dx.doi.org/10.1053/j.gastro.2010.12.006] [PMID: 21147110]

[71]   Haber, R.S.; Rathan, A.; Weiser, K.R.; Pritsker, A.; Itzkowitz, S.H.; Bodian, C.; Slater, G.; Weiss, A.; Burstein, D.E. GLUT1 glucose transporter expression in colorectal carcinoma: a marker for poor prognosis. *Cancer,* **1998**, *83*(1), 34-40.
[http://dx.doi.org/10.1002/(SICI)1097-0142(19980701)83:1<34::AID-CNCR5>3.0.CO;2-E] [PMID: 9655290]

[72]   Graziano, F.; Ruzzo, A.; Giacomini, E.; Ricciardi, T.; Aprile, G.; Loupakis, F.; Lorenzini, P.; Ongaro, E.; Zoratto, F.; Catalano, V.; Sarti, D.; Rulli, E.; Cremolini, C.; De Nictolis, M.; De Maglio, G.; Falcone, A.; Fiorentini, G.; Magnani, M. Glycolysis gene expression analysis and selective metabolic advantage in the clinical progression of colorectal cancer. *Pharmacogenomics J.,* **2017**, *17*(3), 258-264.
[http://dx.doi.org/10.1038/tpj.2016.13] [PMID: 26927284]

[73]   Cui, R.; Shi, X.Y. Expression of pyruvate kinase M2 in human colorectal cancer and its prognostic value. *Int. J. Clin. Exp. Pathol.,* **2015**, *8*(9), 11393-11399.
[PMID: 26617865]

[74]   Notarnicola, M.; Messa, C.; Caruso, M.G. A significant role of lipogenic enzymes in colorectal cancer. *Anticancer Res.,* **2012**, *32*(7), 2585-2590.
[PMID: 22753716]

[75]   Huang, F.; Zhang, Q.; Ma, H.; Lv, Q.; Zhang, T. Expression of glutaminase is upregulated in colorectal cancer and of clinical significance. *Int. J. Clin. Exp. Pathol.,* **2014**, *7*(3), 1093-1100.
[PMID: 24696726]

[76]   Simões, R.V.; Serganova, I.S.; Kruchevsky, N.; Leftin, A.; Shestov, A.A.; Thaler, H.T.; Sukenick, G.; Locasale, J.W.; Blasberg, R.G.; Koutcher, J.A.; Ackerstaff, E. Metabolic plasticity of metastatic breast cancer cells: adaptation to changes in the microenvironment. *Neoplasia,* **2015**, *17*(8), 671-684.
[http://dx.doi.org/10.1016/j.neo.2015.08.005] [PMID: 26408259]

[77]   Gao, XX; Xu, YZ; Zhao, MX; Qi, J; Li, HZ; Wu, JG Progress in nuclear magnetic resonance spectroscopy for early cancer diagnosis *Pectrosc. Spect. Anal.,* **2008**, *28*(8), 1942-1950.
[PMID: 18975839]

[78]   Komoroski, R.A.; Holder, J.C.; Pappas, A.A.; Finkbeiner, A.E. 31P NMR of phospholipid metabolites in prostate cancer and benign prostatic hyperplasia. *Magn. Reson. Med.,* **2011**, *65*(4), 911-913.
[http://dx.doi.org/10.1002/mrm.22677] [PMID: 20967792]

[79]   Placzek, W.J.; Almeida, M.S.; Wüthrich, K. NMR structure and functional characterization of a human cancer-related nucleoside triphosphatase. *J. Mol. Biol.,* **2007**, *367*(3), 788-801.
[http://dx.doi.org/10.1016/j.jmb.2007.01.001] [PMID: 17291528]

SUBJECT INDEX

Atta-ur-Rahman & M. Iqbal Choudhary (Eds.)
All rights reserved-© 2019 Bentham Science Publishers

www.ingramcontent.com/pod-product-compliance
Lightning Source LLC
Chambersburg PA
CBHW041700210326
41598CB00007B/475